PRAISE FOR
ORGANIZE YOUR MIND, ORGANIZE YOUR LIFE

"A treasure trove of tips, tools and techniques, making it possible
to stay mindful of your self-care priorities while navigating
the challenging stresses of everyday life."
—Pam Peeke, MD, MPH, FACP
Host, Discovery Health TV's *Could You Survive?*
Author of *New York Times* bestseller *Fit to Live*
WebMD's Lifestyle Expert

"Marvelous! This empowering collection of transformative, science-
supported tools can help anyone change his or her life in healthier,
happier directions. If you want a smart, straightforward guide
to taming the crazy-making factors in your life, fulfilling
more of your pers...

Editor-i...
Senior Vice Pre... ...e Time Fitness

"Hammerness and Moore have translated the latest science
in brain function into a few, highly effective skills that help
us bring order and control in our lives. In a world where
distractions are ever growing and taking new forms, this book
offers key insights that will help us lead less stressful
and more productive lives at work as well as at home."
—Jon Ayers
Chairman, President & CEO, IDEXX Laboratories

Organize your
MIND

Organize your
LIFE

Organize your MIND

Organize your LIFE

TRAIN YOUR BRAIN
TO GET MORE DONE IN LESS TIME

Paul Hammerness, MD & Margaret Moore

with John Hanc

ORGANIZE YOUR MIND, ORGANIZE YOUR LIFE

ISBN-13: 978-0-373-89244-0

The health advice presented in this book is intended only as an informative resource guide to help you make informed decisions; it is not meant to replace the advice of a physician or to serve as a guide to self-treatment. Always seek competent medical help for any health condition or if there is any question about the appropriateness of a procedure or health recommendation. The medical case studies cited in this book are composites and not based on any one particular individual.

Hammerness, Paul Graves.
Organize your mind, organize your life: train your brain to get more done in less time / Paul Hammerness and Margaret Moore, with John Hanc.
         p. cm.
ISBN 978-0-373-89244-0
1. Distraction (Psychology) 2. Orderliness—Psychological aspects.
3. Time management—Psychological aspects. I. Moore, Margaret, MBA. II. Hanc, John. III. Title.
BF323.D5H36 2012
153.4—dc23
                    2011021307

www.Harlequin.com

Printed in U.S.A.

CONTENTS

HOW ORGANIZED ARE YOU?

(Please answer A, B or C.)

A. VERY ORGANIZED. My desk is neat, I never miss an appointment or a deadline, my friends are amazed, my co-workers are jealous and my boss loves me.

B. MODERATELY ORGANIZED. I manage to stay on top of things pretty well, but sometimes I feel overwhelmed, not sure what to do first, and I must admit that I'm a little jealous of my colleagues and my boss who seem more organized.

C. COMPLETELY DISORGANIZED. In fact, I'll be lucky if I can remember where I parked my car. That's assuming I don't get a text or a phone call in the next two minutes, which will completely throw me off and . . . what was the question again?

If you answered A, B or C, this book is for you! In *Organize Your Mind, Organize Your Life,* we share with you the six key ways in which you can use "top-down organization" to get more done in a lot less time—and feel good about it.

By "top-down organization," we mean brain science. As you will see, there are amazing new insights gleaned about the way our brain works to organize our thoughts, actions and emotions. Through high-tech brain scans, or neuroimaging, we can now "see" the response of the brain to various situations. Here's an exciting example of what scientists have found.

THE ORGANIZED BRAIN IN ACTION

In a 2008 study, subjects were shown a series of pleasant, unpleasant and neutral pictures while they were attempting the difficult task of keeping in check their emotional reactions. Through the use of high-tech brain imaging or neuroimaging, researchers were able to observe the "thinking" regions of the subjects' brains (including areas called the prefrontal cortex and the anterior cingulate cortex) managing the "emotion"-generating parts of the brain. It's an intriguing new study that sheds light into the brain's own built-in system of organization and regulation—one that strives for order, one that can "tamp down" (suppress) our emotions when necessary.

As we will show you, once you can better manage your emotions, you can then begin to harmonize and focus the various "thinking" parts of your brain, opening up a whole new world before you. You're on your way to achieving a more organized, less stressful, more productive and, in many ways, more rewarding life. And—here's the most exciting part—the features in the brain's magnificent self-regulation system come "preloaded" in every functioning human mind; these features can be accessed, initialized and utilized to allow you to become better organized and to feel more on top of things.

You just have to know how to do it. That's what this book will show you.

WHAT MAKES THIS ORGANIZATION BOOK DIFFERENT?

This is not a book meant to give you tips on how to rearrange your desk, to make lists or to set up a better system for keeping track of your appointments.

This is a prescriptive book that will help you better organize your life by better organizing your *mind*, by making some basic changes in the way you think about and deal with your work, your colleagues, your family and yourself on a day-to-day basis. As a result, you will become better focused, more attentive, less distracted and better able to adapt to new situations and changes that, in the past, might have overwhelmed you.

Organize Your Mind, Organize Your Life is organized differently than most self-help books. At its core is a unique partnership between a leading Harvard clinician-researcher and a leader in coaching for health and well-being—a collaboration that serves as a model for the future and can help make a big impact on readers like yourself. In a physician-coach partnership, a new concept in personal health, a Doctor of Medicine diagnoses the problem, explains what you need to do and plants the seeds for you to make the change. Then, a certified wellness coach guides you through implementation of the change.

Here is our team:

Paul Hammerness, MD, is an Assistant Professor of Psychiatry, Harvard Medical School; Assistant Psychiatrist, Department of Psychiatry, Massachusetts General Hospital; and Child and Adolescent Psychiatrist, Newton Wellesley Hospital. Dr. Hammerness has been involved in research on the brain and behavior for the past 10 years, with a focus on Attention Deficit Hyperactivity Disorder (ADHD). He has lectured on the topic locally, nationally and internationally to other physicians, mental health professionals, educators and families. In his clinical practice, Dr. Hammerness sees on a daily basis what a clinically "disorganized" mind looks like across the age spectrum, whether it's an eight-year-old who is struggling in school due to inattention or a forty-eight-year-old professional woman whose life-long organizational problems are now affecting her work and family life. From research, and from witnessing the struggles of people with clinically "disorganized" or distracted brains, Dr. Hammerness shares his insights into what a well-ordered brain can do.

Margaret Moore, aka Coach Meg, is codirector of the Institute of Coaching at McLean Hospital, and a founding advisor of the Institute of Lifestyle Medicine at Spaulding Hospital, both affiliates of Harvard Medical School, founder of a leading coach training school, Wellcoaches, and co-author of a coach training textbook. Margaret and the thousands of coaches she has trained have helped guide tens of thousands of clients through important and positive changes in their health, work and personal lives.

We mentioned the preponderance of books on getting organized that are available. Maybe there are a couple right next to this one. While many of them are good, they often use a bit of an outdated model that begins with organizing your surroundings—your office, your desk, your household—rather than organizing *your mind.* Dr. Hammerness and Coach Meg have a new approach based on the latest scientific literature that employs a top-down (that is, starting with your brain) organizational process—achieved by first understanding six key brain concepts and then employing specific coaching strategies to integrate each of these into your daily life, with astounding results.

These concepts refer to brain or "cognitive" traits and abilities that we all have but that most don't recognize nor know how to utilize. Think of them as embedded features in your brain, waiting to be switched on. Dr. Hammerness will show you where the switch is located and how it works, and Coach Meg will show you how to engage it. So as with the four-wheel drive in your car, you can cruise smoothly over the roughest roads into a more organized and productive future.

These cognitive features can be learned and practiced through the innovative method of self-coaching. They will help you become better organized, less distracted, more focused—with a mind poised and ready to surf the heavy waves of distraction that come rolling in on us in today's world.

To help you become better organized, we have organized this book into a prescriptive "one-two punch" that will enable you to understand clearly the brain science behind these cognitive skills, and then help you adapt it as part of your own make-up.

It's science, followed by solution.

COACHING: THE ORGANIZATIONAL SECRET

That solution—how we will help you to get on top of things, to tap into your "embedded" organizational abilities, improve focus and attention and better structure your life—is one of the unique features of this book. To help you learn how to better function in this distracted world, we will use the new but highly effective psychological technique known as *coaching*, which coauthor Margaret Moore, aka Coach Meg, will explain further in the second chapter. Defined by some as the art and science of facilitating positive change, coaching is essentially a process for developing a road map for wellbeing—and becoming motivated and confident in our ability to implement it.

In this book, Coach Meg approaches the reader as she would one of her clients in practice. Think of her as your coach, a collaborator, helping guide you through the journey of positive change that is the hallmark of what successful coaching is all about. We will take the journey together, and the process begins with what it is that you're feeling—about your emotions, about your sense of organization or lack thereof, about your life.

That's the "one-two" prescriptive punch of this book.

Dr. Hammerness identifies and explains the organizing principles (or, as we call them, our Rules of Order) that are the hallmarks of an attentive, focused brain—one that is able to shift, adapt and function at maximum effectiveness even amidst the constant bombardment of stimulus that is today's world.

Coach Meg shows you how to make these principles your own. She helps you help yourself and guides you step by step toward a more organized mind and, more importantly, toward becoming a better functioning person, enjoying a more productive life.

While their knowledge is rooted in neuroscience, psychology and the science of change that underlines coaching theory, their prescription for you is clear, practical, motivational and—most of all—doable.

You *can* improve your level of organization; you *can* learn to tune out the distractions in your life; you *can* learn to ride the waves of change in a fast-changing world.

Let's go back to that little quiz. If you answered B or C (or even A—because maybe you're rethinking that response as you realize you forgot to reply to the guy from sales who e-mailed you the other day), you are not alone.

By all measures, we are living in a distracted, unfocused world. Call it the flip side of the digital revolution that now gives us such fast access to unlimited amounts of information and that has opened up so many new channels of instant communication. It's great to be able to use Facebook to find your old high school friends, right? It's so convenient to use Google or Bing to find the study you were looking for as opposed to going to a library, isn't it? Can you imagine not being able to send an e-mail to a colleague or a client?

Of course, when all those colleagues and clients e-mail you back and, at the same time, your boss is calling you, and your kids are texting you, and your friends are instant messaging you, well, then you might be forgiven for a bit of nostalgic longing. There was a time when you weren't always so reachable, no matter where you were, no matter the time; and when you weren't always being bombarded by so much stimuli, whether in the form of e-mail, texts, tweets or whatever new technology may emerge... well, any minute now. "'Information overload' has become almost a cliché," writes the Institute for the Future, a think-tank in Palo Alto, California, in a 2010 report on cognitive overload. "We use the phrase half-jokingly to describe the stress associated with the onslaught of media that digital technology has unleashed on us. The sobering reality is that we ain't seen nothing yet. The suffocation of endless incoming e-mail demanding immediate response, the twinge of guilt from falling behind on your RSS feeds, dread about a TiVo hard drive full of unwatched shows—these are all just a teaser for what's to come. No matter how many computers surround us, collecting, aggregating and delivering information, we each have only one pair of eyes and ears, and more importantly, one mind, to deal with the data."

One mind, indeed—but that's where the solution lies.

THE DISTRACTION EPIDEMIC

Nowhere is information overload more evident than in the United States, where some people consider this the psychological equivalent to the obesity epidemic. We even have an unofficial president of Distracted America. No, not the one in the White House but rather in Albany, New York. There, the risks of distraction and disorganization were crystallized in a single, career-flame-out moment in the summer of 2009—a now-infamous moment that made Malcolm Smith a punch line and a punching bag, as well as a cautionary tale.

Smith, a Democrat, was the New York State Senate Majority Leader who famously fiddled with his BlackBerry, checking e-mails, while billionaire Thomas Golisano, a major independent political player in New York, was trying to talk to him. Golisano, who had made a special trip to Albany to meet with Smith, was furious. "When I travel 250 miles to make a case on how to save the state a lot of money and the guy comes into his office and starts playing with his BlackBerry, I was miffed," he told reporters.

Golisano was so miffed that he went to the Republicans and told them he'd be happy to help unseat Smith, perhaps in the hopes of having him replaced with someone who could pay attention for a few minutes. Faster than you can say "you've got mail," the state Republicans engineered a coup, Smith's party was divided, the opposition was poised to take back control of the Senate, and the majority leader was being pilloried in the news media.

"Smith Fiddles with BlackBerry While Senate Burns!" read one headline.

"Blame it on the BlackBerry!" crowed another.

Wrong—blame it on distraction. What cost Smith dearly, and plunged one of the largest states in America into one of the worst constitutional crises in its nearly 235-year history, was (besides maybe some bad manners) a lack of focus, divided attention.

The problem isn't limited to the United States, either. One of the biggest scandals in the British tabloids in 2010—right up there with Duchess of York Sarah Ferguson's admission that she accepted bribes to give business officials access to her influential ex-husband—involved a union official who, during emergency meeting negotiations with British Airway officials hoping to avoid a strike, sent Twitter messages—some at the rate of three or four an hour. When airline officials found out he was tweeting while they were supposed to be talking, they were furious; the negotiations broke down and the strike was on, disrupting travel plans for thousands of people on one of the world's biggest airlines. "Twitter Blamed for Wrecking British Airway Peace Talks," screamed London's *Daily Telegraph* on its front page. Again, the wrong culprit: Twitter is not to blame. Rather, it's a brain unable to stay focused even in a critical meeting that demonstrates an inability to put down a handheld device and look another human in the eye.

Still, at least, Malcolm Smith and the British union official weren't behind the wheel of a car. The National Highway Traffic Safety Administration estimates that at least 25 percent of all auto crashes involve some sort of driver distraction—and there are those who believe this number is steadily climbing as those distractions multiply with the addition of each new mobile communications device, every cell phone feature, every new satellite radio station, every new sign on the road.

But are the signs, the phones and the stations themselves really the problem? Once again, no.

The problem is that we can't deal with them. The problem is that we can't focus. The problem is that we're overwhelmed and disorganized, and the net effect of the Distraction Crisis can be felt in the workplace, at home and in our individual health.

Some other distressing distraction-related statistics:

- Forty-three percent of Americans categorize themselves as disorganized, and 21 percent have missed vital work deadlines. Nearly half say disorganization causes them to work late at least two times each week.[1]

- A lack of time management and discipline while working toward [financial] planners' professional goals contributes to 63 percent of those surveyed facing obstacles regarding their health. There is a direct correlation between too much stress, deteriorating health and poor practice management.[2]

- Forty-eight percent of Americans feel that their lives have become more stressful in the past five years. About one-half of Americans say that stress has a negative impact on both their personal and professional lives. About one-third (31 percent) of employed adults have difficulty managing work and family responsibilities. And over one third (35 percent) cite jobs interfering with their family or personal time as a significant source of stress.[3]

- In a Gallup poll, 80 percent of workers said they feel stress on the job, nearly half said they need help in learning how to manage stress and 42 percent said their coworkers need help coping with stress. Job stress can lead to several problems, including illness and injury for employees, as well as higher insurance costs and lost productivity for employers.[4]

- According to the Centers for Disease Control and Prevention, 80 percent of our medical expenditures are now stress-related.[5]

- Seventy percent of employees work beyond scheduled time and on weekends; more than half cited "self-imposed pressure" as the reason.[6]

One specific category of disorganization or, to be precise, distraction has come to symbolize an era of divided attention: distracted driving. The Department of Transportation (DOT) has a special website dedicated to this problem (distraction.gov), in which readers are reminded about the perils of distracted driving, which is often thought of as just texting but also includes driving while talking on a cell phone, watching a video, reading a map or other behaviors that involve taking your eyes off the road or away from the safe operation of your vehicle.

The scope, effects and consequences of distracted driving are sobering, according to statistics compiled by DOT:

- Using a cell phone while driving, whether it's hand-held or hands-free, delays a driver's reactions as much as having a blood alcohol concentration at the legal limit of .08 percent.[7]

- Driving while using a cell phone reduces the amount of brain activity associated with driving by 37 percent.[8]

- Nearly six thousand people died in 2008 in crashes involving a distracted driver and more than half a million were injured.[9]

- The younger, inexperienced drivers under twenty years old have the highest proportion of distraction-related fatal crashes.

- Drivers who use hand-held devices are four times as likely to get into crashes serious enough to injure themselves.[10]

Lest we assume, as many seem to do, that distracted driving is purely a problem of the young; teenagers and young adults who are checking their friends' Facebook status while doing ninety miles per hour on the interstate, think again: almost half of adults who send text messages have sent them while driving, according to a 2010 study by the Pew Research Center (the same study found that about one-third of sixteen- and seventeen-year-olds admitted that they had done the same). According to distraction.gov, half of all people in the United States admit to cell phone use while driving; one in every seven admit to sending cell phone text messages while driving. These are also folks who should know better: 65 percent of drivers with a higher education text or talk while driving.

All in all, the distracted driving crisis—part of that larger Distraction Epidemic—seems to some a part of an even greater problem, suggesting that the human race has reached a point of information overload—or at least a point where we feel so overwhelmed by the demands of our lives that we would risk our lives for one more text or phone call. In 2010, *The New York Times* published a series of articles about the supposedly dire effects of technology on our brain. In a *USA TODAY* story on the issue, one researcher concluded gloomily that "people are multi-tasking probably beyond our cognitive limits."

A DISTRACTED FACT OF LIFE?

Some say there's little that can be done about all of this. The pace of life is increasing and the distractions multiplying. Get used to it. You're powerless. To which we say, baloney! While we may not be able to slow down technological change or the speed with which life unfolds around us—and in some cases, why would we want to?—we very definitely can find a way to better manage ourselves, in order to not only deal with change and complexity but also thrive amidst it. This book is designed to show you how.

Remember: for every driver driven to distraction and for every stressed-out person who has lost an assignment, a job or a vital piece of information because he or she was disorganized and distracted, there are people on the opposite end of the spectrum. These are individuals who know how to use their brain's abilities to organize their lives, to stay focused on the tasks at hand and to enjoy greater productivity—and pleasure!—at work and at home.

Some of them you probably know: athletes such as Derek Jeter or Tom Brady, famous for their ability to block out distraction and focus on the little white ball or the white line on the field ahead, public servants such as General David Petraeus, making life-and-death decisions in the midst of a foreign country exploding in religious civil war; Steve Jobs, a visionary who manages one of the world's largest and most influential corporations; Hillary Clinton, patiently mastering the minutiae and intricacies of a seemingly intractable conflict as she engages Palestinians and Israelis at the bargaining table. And the ranks of the super-organized are not limited to government, big business or the pressure cooker of professional sports: how about J.K. Rowling, whose disciplined imagination enabled her to create the Harry Potter world? (Imagine how organized she had to be to keep track of, much less create, the Hogwarts faculty and their complex histories.)

There are numerous examples of famous people whose achievements lie, at least to some degree, in their ability to stay calm, focused and organized, especially in the midst of crisis. There are many other very successful people whose names might not make headlines but who have, through both innate and learned skills, managed to harness their cognitive powers in a way that makes them extraordinarily productive, both on the job and at home.

Let's meet two of them.

ORGANIZED MINDS AT WORK AND PLAY

By 8:30 am most mornings, Rob Shmerling has already exercised for an hour, has caught up on world and national news, and is well into responding to his e-mails.

For two hours, he exchanges messages with colleagues and scours various websites for the latest medical news. Dr. Shmerling is a physician and the clinical chief of the division of rheumatology at Beth Israel Deaconess Medical Center in Boston.

It's a big administrative job at one of the country's leading hospitals—but it's not all that he does. Shmerling, fifty-four, also writes and does research—he has authored a total of forty-one journal articles, book chapters or reviews, as well as numerous web stories for nonexpert audiences. He also teaches and mentors medical students and residents. He is a husband and a father of two daughters. He volunteers at a women's shelter once a week. He and his wife belong to a book club (Janice Y.K. Lee's *The Piano Teacher* and Kathryn Stockett's *The Help* are two recent novels they've enjoyed). He also "hacks away" at the piano, is an amateur photographer and, on weekends, enjoys long bike rides in the Massachusetts countryside.

Oh, and he washes and folds socks, too.

"I'm the laundry guy," he says proudly. "Everybody in our house has their job, and that one's mine."

Actually it's one of many jobs, as you can see.

How does Shmerling cram it all into one day, one week, one *life*, and make it look easy?

He admits that he is a creature of habit and was always fairly structured. "I can recall organizing the crayons by color in those sixty-four-Crayola packs as a little kid," he says with a laugh. But, he's quick to add, a lot of the skills that help keep him organized he learned because he *had* to. And he's still learning. "I've gotten better at ignoring things," he said. For example, "We have this e-mail system where a quick preview of the e-mail comes up on your screen, and at first it was distracting. Now I've gotten better at sticking with the matter at hand. If it's a really important message, I can attend to it, but I don't let them distract me as they pop up."

In the hospital, things come at Dr. Shmerling fast and furious. A patient's condition might change. An administrative problem may arise. A resident or a nurse or a colleague may need an immediate answer. And sometimes the decisions really are a matter of life and death. "I used to get more easily flustered when several things were coming at me," he says. "Now I've learned how to deal with it. Now I can shift pretty quickly from one thing to another and prioritize."

The problems that do come up are often complex ones—what course of action to prescribe to someone with arthritis, lupus or osteoporosis; dealing with patient complaints or concerns; helping to mediate or referee internal problems that arise, whether with staff or fellow physicians. He knows how to act, but he also knows how to think *before* he acts.

"I try to imagine the range of options for a given situation and figure out fairly quickly if this is something I've seen before," he explains. "If not, if it's something better done by someone else, or if I'm going to need someone else's help solving this, I mentally file it away, putting it aside for later."

Putting his attention on and pulling it off, deftly and smoothly, as the need arises—that's a sign, as we'll see, of an organized mind. Dr. Shmerling does it with a range of tools, some high-tech, some not. "If I have to jump off something, I'll bookmark what I was working on," he says. "Either with a mental or actual Post-it note so I can return to the right place quickly later on." He also has a nice trick for keeping track of his reading (and in his job, he does a lot of it—reports, memos, articles). If he's reading a Word document on the computer, "I'll yellow-highlight the line I'm on so I can get right back to the page and the line I was on, without wasting time scanning through the document, going 'where was I?'"

Shmerling uses a PalmPilot to keep track of appointments and to have other important information at a glance when he needs it, even though, he admits, "I'm regularly laughed at for using a device so ancient." And while you might think someone being held up as an exemplar of efficient organization would have an empty, ordered desk at the end of each day, it's not the case. Dr. Shmerling's offices at home and at the hospital are filled with stacks of books and papers—but, he says, "While it might not look organized to you, I know exactly where everything is."

The efficiency allows him some simple pleasures during the work day. People who feel overworked often claim they have no time to read anything but e-mails or work-related documents. Shmerling not only finds time to read *The Boston Globe* every morning online, he spends an extra few minutes doing the popular Sudoku numbers puzzle; and is a diligent fan of *Doonesbury* and *Dilbert* ("Another efficient office guy!" he jokes). Indeed, while he is a hard-working professional and leads a busy life, Dr. Shmerling is not some obsessed workaholic, constantly looking to squeeze another hour out of his life to devote to work. He likes to have fun, he likes to laugh, he has a rich and satisfying personal life and, oh yes, some of that time he manages to save by being efficient and organized, he likes to waste.

Here's an example: "I like to stop sometimes on my way to work and have Starbucks. If I was really trying to be a time management-efficiency nut, I could save a few minutes by making the coffee at home or grabbing it at the hospital cafeteria. But I like stopping at the coffee shop. It makes the ride more pleasant. Nothing wrong with a little down time."

A graduate of Harvard Medical School, Shmerling is obviously a smart guy. But he is quick to point out that his academic pedigree has nothing whatsoever to do with his ability to be efficient. "There's nothing I learned at Harvard or anywhere else specifically that taught me any of this," he says. "None of it requires any particular advanced degree. The measures I take to keep organized could certainly be adopted by others."

Some of those are common sense and can be found in any of the dozens of books about organization. "Make a list of what you need to do tomorrow, at the end of each day." Fine. Good tip. But there's more at play here. The skills that Shmerling demonstrates—his ability to shift from one problem or stimulus to another, to sustain his focus, to attend to several things at once while prioritizing quickly the one that is most demanding of his attention and to do it with ease and grace while maintaining composure and good humor—speak to qualities that are linked not to the layout of his office but the make-up of his mind.

It's an organized mind and, while he may have certainly nurtured it, nature created it that way. We all have the systems, the functions in our mind that enable us to become better organized, whether our job involves, as Dr. Shmerling's does, people's lives—or our life savings, as is the case with our next organized role model.

Let's take a peek at a typical day for another organized person.

Catherine Smith starts her morning on the roads and trails outside her Connecticut home. Her daily, three-mile run is not only good for her heart but also her head. "It's a re-energizing time," she says, "but I also use it to clear my head, to think and to reflect." Thinking! What a concept. Who has the time? But using that time to plan and reflect may be one of the keys to Smith's success—and quite a success she has been. Until recently, she was one of the highest ranking female business executives in the global operations of ING Insurance. Headquartered in Amsterdam, ING is one of the world's largest insurance and financial services corporations. Smith was CEO of the division that oversees workplace retirement plans in the United States. She managed a business that employs about 2,500 people and serves nearly 5.5 million consumers at more than 50,000 private, public and nonprofit employers throughout the country. (You may very well have your retirement money in the division of ING that Smith oversaw.) Their combined assets today: a staggering $300 billion, literally more than the gross national product of many countries. And it was her responsibility.

Smith was accustomed to traveling one out of every two working days. On a daily basis she made decisions that involved millions of dollars—many of them representing people's life savings and retirement money. "ING is doing important things," she acknowledged. But she had fun doing it. "I have a lot of passion and energy," she says when asked how she managed to stay on top of everything she needed to do. (She has since taken this passion and energy to a brand new role—and one no less demanding—serving the state of Connecticut as Commissioner of Economic and Community Development, a position to which she was appointed by the state's governor, Dannel P. Malloy.)

A former colleague who traveled with her on a daylong business trip in New England while at ING commented admiringly in an e-mail how effortlessly Smith seemed able to meet all of the demands and responsibilities hurled at her:

> *Early morning:* in Quincy, MA, visited one of the company's major sites
>
> *Late morning:* in car on the way back to Hartford area— did a phone interview with major trade publication
>
> *Noon:* arrived at golf course in Bloomfield to play in an LPGA tournament that ING sponsored. Won longest drive contest!!
>
> *Evening:* after her gold round, came in and spoke to the crowd about ING's commitment to community and its role as a good corporate citizen
>
> *Late evening:* caught up on e-mails

In addition to her innate talents, she has a mind that is fully engaged, a mind that is organized.

In her new job, she adds, she's putting it to good use.

"Organization is even more important in this role!" says Smith, whose job includes helping to create jobs and attract new business to the state. "It's requiring me more than ever to utilize good time management skills."

Interesting point: Smith doesn't make to-do lists, a supposedly common trait among organized people. She does make the most of her greatest resource, which is between the ears. "I use my reflective time to consider what things I got done, what things I need to do," she says. Smith has also learned how to put aside things and return to them at a more opportune time. These could be complex problems or problem people. Like we all do sometimes, she can get frustrated or angry. The difference is that she knows how to manage those emotions. "It's better to wait until you can speak thoughtfully and calmly," she says. "I'll leave that part of my work alone for a day or two, to get perspective and calm down."

This reveals another part of her cognitive make-up: a mental nimbleness that allows her to jump off of one task and onto another without losing balance. "It's rare that I go through a full day without some interruptions and changed priorities," she says. "You cannot ignore many of these issues and need to be flexible in addressing them." Another thing about Smith: while many might hail her as a paradigm of "multitasking" or as a "juggler," she rejects that very terminology. "I try very hard not to multitask," she says. "Instead, if I can stay focused on the task at hand I find I'm much more effective in completing it. If I try to spread my energies among several things simultaneously, more often than not, I end up with several half-done things." Again, as in the case with Dr. Shmerling, it is not necessarily a driven mind or a person so single-minded that he or she is an automaton, bereft of joy and focused only on work or success. Catherine Smith, too, enjoys what by any definition would be considered a well-rounded, balanced and satisfying life. She has been married to the same man for twenty-seven years, and they've raised two happy and healthy children. She is a passionate outdoorswoman, who enjoys biking and hiking, and also is active in various volunteer and environmental causes. She is on the board of directors of Outward Bound USA (which serves 70,000 students and teachers annually) as well as a former director of the Connecticut Fund for the Environment.

Balance. Flexibility. Poise. An ability to tamp down the emotions and to shift and set your attention on something else with grace and ease.

As we shall see, these are all qualities of the well-ordered mind. That is, a mind that is organized and can focus and pay attention. A mind that can stay afloat and buoyant in a turbulent sea of change.

It's a mind, or a mind-set, that can be yours as well. While you may not have the academic pedigree of Dr. Shmerling or the business resume of Catherine Smith, you do have the capacity to engage and enhance the same cognitive skills that can improve your life. Whether your goals are simply to better focus on your required reading for school or work, better manage your day in order to have more time for your spouse and children or make a quantum leap forward in your career, the ability is there in your mind and in the resources that exist in you, like unused features in your computer that you have but may simply not know how to use.

In the next chapter, Dr. Hammerness will explain the principles—or Rules of Order—and the science behind them by using some cases from his own practice.

In Chapter 2, Coach Meg will show you how to get ready to take the journey of change.

In subsequent chapters, they will examine each of the Rules of Order in depth, giving you both the science behind it—so you have a better appreciation of just how organized your brain is (although you might not feel that way at the moment)—and specific suggestions on how to integrate each of these organizing principles into *your* life.

Citizens of Distracted America! Men and women all over the disorganized world! Join us in becoming more focused and productive. You have nothing to lose but your car keys, which, by the way, you probably left on the kitchen table.

CHAPTER 1

The Rules of Order / *Dr. Hammerness*

It was a Thursday, around 6:00 pm, and I was sitting in my office in Cambridge, Massachusetts, located along a tree-lined stretch of Alewife Brook Parkway, a few miles outside of Harvard Square.

The four-story brick building, an annex of Massachusetts General Hospital's psychiatry department, is where I see patients as part of my research and teaching responsibilities at Harvard Medical School. They span the age and occupation spectrum—elementary-school children, grandparents, lawyers, salesmen, housewives and househusbands—but they have one thing in common: they are coming to see me and my colleagues with familiar complaints and concerns. "I know I could be doing better" is a common one; as is, "I can't go on like this."

While the complaints may vary slightly, the symptoms they describe are the same—and consistent with the condition we treat. You've probably heard of it: attention deficit/hyperactivity disorder (ADHD).

One of those patients, we'll call her Jill, is late for her appointment.

As I sit catching up on e-mails, the door bursts open and in she flies, out of breath from climbing the two flights of stairs to my second-floor office. She is flustered and clearly upset.

"Sorry I'm late!" Jill says, as she plops down on the chair facing my desk. "You wouldn't believe my day."

"Try me," I say. "Take a deep breath and tell me what's going on."

Jill is in her late thirties and a highly educated research scientist, one of the many "knowledge workers" who labor in Cambridge, home of Harvard and Massachusetts Institute of Technology. She takes a moment and launches into her story, which begins a few weeks earlier when she temporarily moved into a friend's apartment while her own house was being renovated.

"Last night, when I came in," she says, "I put my keys down somewhere, and this morning, I had not a clue where they could be."

I nod. I have a feeling I know where this is going.

"I looked everywhere—the usual places, which of course are not the usual places, as it's not my place. My friend, she really is a good friend, but I am wondering if she has more trouble than I do. You think I am disorganized, you should see her place...."

I know this is the right time to jump in and direct our conversation back to the issue at hand or—like this morning—Jill could continue running in verbal circles and not getting anywhere. "Okay, so, you were looking for your keys...?" Jill smiles. "Oh, right, yes, I was flipping out. I spent thirty minutes trying to find my car keys."

Jill then stops, shaking her head.

"Well, did you find them?" I ask.

She nods ruefully. "Eventually."

"Where were they?"

"Right on my friend's kitchen table! And, of course, I'd walked back and forth through the kitchen ten times while I was looking for them. All that time they were right there…right there in front of me. Unbelievable!"

"Sounds very frustrating…but pretty believable, as those keys have eluded you before." Jill smiles ruefully, and I press on. "Then what happened?"

"My day was in shambles from that point on." Jill went on to relate how the half hour she'd spent looking for the keys set off a domino effect of tardiness and inefficiency—problems galore. She arrived at work late for a meeting and opened the door to the conference room just in time to interrupt an important point that one of her company's head honchos was making. Embarrassed and angry at herself, she returned from the meeting and finally got in front of her computer to find a barrage of e-mail reminders that further annoyed and overwhelmed her. She sent out a flurry of responses, including a snippy reply to the wrong person, who was not happy to get it (neither was the correct recipient, when she eventually cleared up the mistake). Dealing with her e-mail gaffe kept her from attending to a project due by noon. Her deadline blown, she skipped lunch, scrambling to get her work done, and what she did hand in—two hours late—was subpar and received with something less than an enthusiastic response by her supervisor.

In other words, it was a crummy day for Jill. It wasn't the first time such a day had begun with something misplaced or by an episode of forgetfulness, but the snowball effect of losing her keys still surprised and upset her.

"This happens all the time," Jill says, teary-eyed, angry and ashamed. "At this rate, I could lose my job…just because I can't keep track of stupid things like keys."

I'm sorry to hear that Jill is upset, but her story is not unusual. Jill has ADHD—and she is certainly not alone. It's estimated that about 4 percent of adults and 5–7 percent of children in this country meet the medical criteria for ADHD. It's equally safe to estimate that at some point in their lives almost everyone has *felt* as if they have ADHD, too. The symptoms of ADHD include forgetfulness, impulsiveness, losing items, making careless errors, being easily distracted and lacking focus. Who hasn't exhibited one of these symptoms in the last few days… or even hours? Who hasn't lost their car keys? Who hasn't been distracted in the car (once the keys are located), on the job or at home—by a text, a tweet, an e-mail, a cell phone ring? Who hasn't been late for a meeting or missed a deadline or made a mistake because they were disorganized that day, lost focus that morning or were distracted that minute? That doesn't necessarily mean you have ADHD, but it does suggest you might be part of the distracted masses that now make up such a large part of our society. If so, you've come to the right place because we're going to show you how to get back on track.

ADHD or OBLT?
(Overwhelmed By Life Today)

If you answer Often or Very Often (on a ranking scale of Never, Rarely, Sometimes, Often or Very Often) to four or more of the following questions, it may be beneficial to consult with a health professional to see if you have ADHD.

In the last six months….

1. How often do you have trouble wrapping up the final details of a project once the challenging parts have been done? *(never/ rarely/sometimes/often/very often)*

2. How often do you have difficulty getting things in order when you have to do a task that requires organization?

3. How often do you have problems remembering appointments or obligations?

4. When you have a task that requires a lot of thought, how often do you avoid or delay getting started?

5. How often do you fidget or squirm with your hands or feet when you have to sit for a long time?

6. How often do you feel overly active and compelled to do things, like you were driven by a motor?

Source: World Health Organization

Whether or not you have ADHD—and chances are, you probably don't—the purpose of this book is to inform, inspire and organize your brain. Whether forgetfulness is a "symptom" of a disorder for a person like Jill or an "issue" for someone else who doesn't have the same degree of severity, this book will approach it in a straightforward way—and with equally straightforward and effective solutions.

What was first labeled the "Distraction Epidemic" by *Slate* magazine in 2005 has now reached epic proportions, right up there with the obesity epidemic and is of no less import than that or other public health crises that have befallen modern society. In a 2009 *New York* magazine story on the attention crisis, David Meyer of the University of Michigan described it as nothing less than "a cognitive plague that has the potential to wipe out an entire generation of focused and productive thought" and has drawn comparisons to the insidious damage of nicotine addiction.

"People aren't aware of what's happening to their mental processes," says Meyer, "in the same way that people years ago couldn't look into their lungs and see the residual deposits." The difference here is that unlike the "mad men" of the 1950s and 1960s who went around merrily sucking up packs of unfiltered Camels, seemingly oblivious to the harmful effects, most of us today know that we are having problems staying focused, paying attention and maintaining some sense of order in our lives.

Unlike smoking (which you either do or don't do), it's not just the people afflicted by the most serious and definable form of distraction and disorganization—ADHD—who are affected by this epidemic. Ask friends, family members and colleagues how they're doing, and chances are, the responses will usually include words like "frazzled," "stressed," "overwhelmed" and "trying to keep my head above water." In casual conversation, you often hear people talking about "brain freezes," "blanking out" on something or suffering "senior moments" (often, when they really aren't very senior). All of them…all of us…are affected to some degree by the epidemic.

To get back to my patient Jill in the four-story brick building in Cambridge, Massachusetts, I knew that the woman with the lost keys and the lousy day was not one of the millions complaining to each other about how crazed their lives have become. She has a clinical disorder; most do not. But, as I listened to Jill's story, I also knew the potential power of a rather simple solution that could help her and many others.

A couple of weeks earlier, during one of our regular sessions, Jill and I had somehow gotten on to the topic of the Apollo lunar landing. We talked about the coverage of the fortieth anniversary of that historic moment, the spectacle of the great Saturn rocket that hurled the astronauts into space, how exciting it still was to see the old black-and-white images of Aldrin and Armstrong on the moon and hear their voices crackling over the television from Tranquility Base and about whether we'd ever go back.

The memory of that conversation about the space program and her interest in it gave me the language needed to help frame the solution for Jill.

"So, I have a thought about how to start your day tomorrow," I say. "As we've been talking about, we are working on bringing order into your life, changing old patterns that don't work with new ones that do."

"Right, that sounds good," she says attentively. "What's your idea?"

"You need a *launch pad* for your keys."

Her eyebrows raise quizzically.

"A launch pad," I repeat. "A place where you always put your keys and maybe your ID and glasses, too. That way, you'll know that's the place they're always going to be…and every morning, that's where you'll launch your day."

Slowly, as if an unseen hand was drawing it methodically, a smile etches itself across her face.

"A launch pad," Jill says, starry-eyed "Yes, a launch pad. What do I have to use? A box…a hook…a basket…a tray?"

I smile back. "It's *your* launch pad. You can use whatever you like. You just need to make sure you know where it is and keep it in the same place…so that the moment you enter your friend's house, you'll leave your keys there and then every morning that's where they'll be. On the launch pad, ready to lift off."

This seemed to really resonate with Jill. First of all, it was an action-oriented solution, something she could do right away and without great difficulty. But more importantly, and Jill appreciated this, the launch pad served as an image, a reminder of how one's day can begin, not in confusion and distraction but with precision and predictability.

The next week, Jill arrived for her appointment on time. And she entered the room not in a huff but with a smile.

"Go ahead," she says, "ask me about my forgetfulness this week. I'm ready to answer."

"Okay," I respond. "So tell me, did you forget any items, appointments, things like that this past week?"

"Nope," she said triumphantly, "and here's why." She reached into her pocketbook and pulled out a small, uncovered trinket box, one, she explained, that she hadn't used in years. "My launch pad," she says, proudly. "I have a spot for it right by the kitchen door." Moreover, Jill went on to tell me, she had not neglected the area around the launch pad. In fact, you could say that a major redevelopment project had been undertaken in the area: the table cleared and the space near the door rearranged so that her launch pad had its own...well, *space*. That wasn't all, she reported. She built a launch pad at her office, too—but this one was project oriented for critical tasks to distribute to others. This, too, was accompanied by a cleaning and rearranging of her workspace.

That week, you might say, all systems were go for Jill. Is this an ADHD "cure"? No, but it's a small success to build upon. And she has. You could see the impact on her organization and on her self-esteem; she began to regain confidence, as she could now trust herself that her mornings would be a little less frantic and a little more consistent. I'm happy to add that since she "launched" her launch pad, she has not missed a morning meeting again because of time spent looking for her keys.

My experience with Jill illustrates a few important points about organization. First, individual moments of forgetfulness and disorganization can have major consequences.

Second, just as one episode of forgetfulness triggered a series of negative events, so can one small step lead to giant leaps of improvement in the organization of one's life. The launch pad is a simple solution, but it has effects that go far beyond knowing where your keys are. You begin to think about other things you can organize. You have more time. You are less stressed before you leave the house in the morning. You enter a new environment more relaxed and thinking more clearly. And so on and so on.

Third, and this might not be something readily apparent from hearing the conversation with Jill, the simple remedy that I suggested is rooted in an understanding of the workings of the most complex organ known: the human brain.

THE ORGANIZED BRAIN: TAKE A LOOK

You may have heard about how neuroimaging—our ability to look at the structures and functions of the working brain through advanced imaging technologies—is giving us incredible insights into our understanding of how the mind works. That's true, and nowhere more so than in our ability to see how the brain is structured to help it function optimally—in other words, its organization.

So just how is the brain organized? Well, at first glance, its complexity seems almost beyond comprehension. The human brain is composed of neural cells—an estimated 100 billion neurons!—that are connected into groups or circuits, communicating with chemicals called neurotransmitters. These groups form larger macrocircuits. The scale of it all is mind-boggling. But here's a good way to visualize it: Think about looking at your house on Google Earth. You can zoom in and see where you live and your neighbors' houses—each of them like a single neuron. Toggle back on your computer, and you can see a whole block. Go back further, and the blocks form a neighborhood, a community. Even further, and you're at jet-plane level, looking at clusters of communities forming a metropolitan area. The brain is structured in a similar way. Put all those individual "houses" (neurons) together, and you go from something relatively simple into something enormously large and complex.

Now imagine it's a hot summer day in your neighborhood, and you and everybody on the block cranks up the air conditioning. Folks on the adjoining blocks are doing the same. If the whole community and the adjacent communities are doing it, too—responding to the hot weather—what do we have? An overload, maybe at the local level, but more likely—if enough blocks or neighborhoods are involved—a grid failure, a blackout, an entire community powerless.

What happened is that the system got overloaded. But it probably could have been avoided. Chances are, there were warnings signs: The lights dimmed at one point. Or the local power authority issued alerts throughout that day, warning customers to cut back on their power usage during peak hours.

A brain bombarded with too much stimulus, as many of us are these days, is similar to the community on the brink of a power outage on a hot summer's day. Too much drain, too much strain. Losing those keys, forgetting a scheduled meeting, "blanking out" something you needed to do: each of these episodes are like a momentary dimming of your cognitive lights, a warning message from the brain. Indeed, you may have already experienced some of these signs, which is why you picked up this book.

That's a great first step. But here's where the electrical blackout analogy falters. There is only so much power available from the grid and when it goes down, it goes *down*. Fortunately, the brain is more adaptable, so we reach for a different metaphor:

You may get irked and frustrated by what goes on in Washington, D.C., but one thing that continually works and works well is the balance of power in our American system of government. The Executive Branch, Congress, the Supreme Court—sure, they may bicker and they may even work against each other at times, but the truth is that in the complex array of checks and balances that is the genius of the Constitution, none can ever get the "upper hand" over the long haul. The human brain, too, is in and of itself a remarkable system of checks and balances of "on" and "off" switches. What's really remarkable is how, despite this delicately engineered balance, the entire structure stands strong and stable, even when being battered by the storms of stimuli that assail us in modern life.

A NEW APPROACH TO NEUROSCIENCE AND MENTAL HEALTH?

A provocative new way of thinking about neuroscience and mental health comes from the folks at the National Institute of Mental Health, who suggest that many cognitive, emotional and behavioral problems—e.g., ADHD, depression, anxiety disorders—can be thought of as problems in the brain's circuitry, problems that may have existed and been ignored for years. If we can identify them early, we may be able to intervene in very specific ways to prevent and even reverse the problem; much the way a physician will prescribe a low-fat diet and exercise to a patient with slightly elevated cholesterol which, if left on its own, can lead to very serious heart and blood vessel problems or failure.

As neuroscience shows us the intricate details of these circuits, we see the brain's checks and balances in action. One example of particular importance at the "macro" circuit level can be seen in the brain's balance of emotions and cognitions. Remember the brain-imaging study that we mentioned in the introduction, the one where subjects viewed pleasant, unpleasant and neutral pictures while attempting to keep in check their emotional reactions? Through the use of imaging techniques, researchers at the University of Colorado were able to observe the "thinking"-brain regions of these subjects (including areas called the prefrontal cortex and the anterior cingulate cortex) actually regulating the emotion-generating regions. If you can manage your emotions, harmonize and focus the various "thinking" parts of your brain, then a whole new world opens up before you. You've got a more organized, less stressful, more productive and, in many ways, more rewarding life—not to mention one where you can always find your car keys.

Yes, this is the good news about your brain. While *you* may be disorganized, your brain isn't. Inherently, it's a jewel of organization and structure, of different components working harmoniously together. And here's the exciting part—the features in this magnificent self-regulation system that come "pre-loaded" in every functioning human mind can be accessed, initialized and used to become better organized and to feel more on top of things.

You just have to know how to do it.

That is the purpose of this book: to help you do for yourself what I did for Jill; to help you understand just what your brain can do to help maintain order and to keep you focused and then to show you how you can do that for yourself. We'll talk big picture and sharp-focus details. We'll talk about day-to-day details, but we'll also talk about life in general. We'll talk descriptive and prescriptive. We'll talk "neuroscience"—the science of cognition, the science of ADHD and the science of a properly functioning brain. And we'll talk "solution"— how you can learn to harness those amazing organizational abilities embedded in our minds. My colleague and coauthor Margaret Moore will also employ an exciting new discipline, the science of change, to help you make these modifications in your life (more about that in the next chapter).

What we will not do, sorry to say, is eliminate distractions. The bad news on that front is that they're here to stay. And some of the things that distract us are very odd indeed.

The Brain Bone's Connected to the Ham Bone…

The issue of distracted driving has been in the news over the past few years. First cell phones and now texting have been shown to be contributing factors in many incidents of distracted or inattentive driving. But you can't just blame technology here. *The Record,* a newspaper covering the Waterloo region of Ontario, Canada, analyzed more than four hundred local highway reports of distracted-driving collisions to see what was causing drivers to take their eyes off the road. Here's what the reporters on *The Record* found:

About 20 percent of drivers were distracted by something inside their vehicle—fiddling with the radio or talking to other passengers.

One driver told police he was driving with his knees while trying to roll up his window. He slid onto the shoulder and smashed into the concrete median.

A passenger told police she was having a heated argument with her boyfriend, the driver. Neither noticed their car had slid onto the shoulder until she grabbed the wheel, causing them to lose control.

Six drivers were distracted by food. One driver admitted she was cleaning melted candy off her steering wheel when she lost control of her car. Another started choking on coffee, and another let go of the steering wheel after spilling hot chocolate.

But in almost half the cases, drivers were distracted by something outside the vehicle, most often other drivers, accidents, construction crews or road signs.

One driver became so transfixed by pigs being transported in the next lane that she crashed her car into the truck.

"As I was in the turn, I looked off to my right at a transport truck in the right-hand lane," she told police in her driver statement. "It looked like he was transporting pigs, so I focused on the animals. As I did, I started to head toward the truck . . . I remember slamming on the brakes. Everything went white and then I heard the crash."

Disclaimer to readers of this book: If you are someone who becomes transfixed by the sight of farm animals in trucks while driving, nothing we can say will help you.

THE RULES OF ORDER

Through years of working with patients, through the growing body of clinical literature and through insights gleaned from advances in neurosciences, we have learned much about what ADHD patients and the general public struggle with. From that, we can better understand what we should do in order to stop being forgetful, start getting focused and stop allowing distractions and a lack of focus to mess up our lives. In *Organize Your Mind, Organize Your Life* we boil down many essential "brain functions" to six principles—what we call the Rules of Order. Consider these "brain skills" or abilities that you can develop and master. In the chapters ahead, Coach Meg and I explain these Rules of Order and then show you how to learn these skills to give yourself more focus and your life greater order. We will start with three "simple" principles and build upon these more complex organizational abilities and strategies.

1. Tame the Frenzy: Before we can engage the mind, we must control, or at least have a handle on, the emotions. It's hard to be thoughtful or efficient when you're irritated, frustrated and distraught. First, it's

necessary to calm down and stabilize the frustrations, anger or disappointments that we may be feeling at that particular moment.

A wonderful example of this quality comes from, of all places, a well-known cable television program. There is no one better at taming frenzy than Cesar Millan from *Dog Whisperer with Cesar Millan*. And just as Cesar teaches dogs and owners how to more happily coexist, so, too, can he teach us something about the necessary approach to thinking and organization. When he deals with dogs (and their often-distraught owners), Cesar's tenet is to be "calm, yet assertive." In order to have a healthy, responsive canine, you have to find your "calm-assertive" energy. As described on his website (www.cesarsway.com), this is "the energy you project to show your dog you are the calm and assertive pack leader." Assertive, he adds, "does not mean angry or aggressive. Calm-assertive means always compassionate, but quietly in control."

Quietly in control. That's a nice phrase. How does this apply to your life and to your abilities to better organize yourself? Here's how: before you attack that mound of work piled up on your desk or computer inbox, you can't be angry over the fact that it's there, annoyed with your boss, fearful of what's ahead or full of self-criticism for letting it get this way again. First, you need to get yourself together, get ready to mobilize your cognitive resources—*then* you can tame the wild pile, like Cesar tames the unruly canine. Organized, efficient people are able to acknowledge their emotions. But unlike many who let their emotions get the better of them, these folks have the ability to put the frustrations and anger aside, almost literally, and get focused on work. The sooner the emotional frenzy welling within you is tamed, the sooner the work is done and the better you feel.

Like Cesar says: quiet confidence.

2. Sustain Attention: Sustained focus or attention is a fundamental building block of organized behavior. You need to be able to maintain your focus and successfully ignore the many distractions around you in order to plan and coordinate behaviors, to be organized and to accomplish something.

In the process of sustaining attention, your brain scans the environment, directing your attention on a certain stimulus, while it continues to process other auditory and visual information. So while your attention rests on one thing (the speaker at the head of the conference table, for example, talking about an important new development at your company), your brain continues to evaluate new information (the rustle of papers to your left, the whispered comment to your right). This extraneous information (or "noise") is competing for your attention, but the organized brain is able to instantly evaluate and screen out what is not worthy of your attention—to identify the signal through the noise. The sound of the papers and the side conversations are deemed unworthy of greater cognitive effort, but the person who rushes into the meeting saying, "Our CEO has just been led out of the building in handcuffs!" would go right to the top of the "Pay Attention!" list.

The ability to properly handle all the noise from the environment— and to evaluate and prioritize it while not being pulled off the main task at hand—is another basic and important sign of the organized brain.

3. Apply the Brakes: The organized brain must be able to inhibit or stop an action or a thought, just as surely as a good pair of brakes brings your car to a halt at a stop light or when someone cuts suddenly into your lane. People who don't do this well struggle with suppressing what has turned out to be the wrong response or action. Often, it is very difficult at times to stop yourself in the middle of something. Here's an example:

You're working diligently on one task—say, your taxes. You're sustaining your focus as you itemize your deductions and carefully read the forms. Meanwhile you've been subjected to an ongoing stream of distractions. Your spouse wants to know where you left the television remote. Your child has a homework problem. A coworker texts you with a question. Then, the phone rings. It's your accountant, calling to ask for a meeting to go over your taxes. Your instinct is to forge ahead, because you really want to finish this tonight so you can watch your favorite television show, which is on tomorrow.

The organized brain says, "Stop now and schedule the meeting!" Yes, it would be easier and more convenient for you to just get it done now. But the organized brain has weighed the options. The organized brain remembers that last year you made a mistake on your tax forms and ended up paying $1,000 (not to mention $500 to your accountant, who had to redo everything). So the organized brain decides to put on the brakes. The function is called "inhibitory control," and you could also think of it as a compassionate hand on the shoulder, or a sort of impulse control that keeps the efficient organized brain from getting off task and helps put you into a position for the next Rule of Order.

However you look at it—traffic cop holding up a raised hand or guidance counselor gently steering you away from an ill-advised task— you need to heed the message of the organized brain and stop in order to get to the next step.

4. Mold Information: Your brain has the remarkable ability to hold information it has focused upon, analyze this information, process it and use it to guide future behavior—even after the information is completely out of sight. This form of brain work involves something called *representational thinking.*

Efficient and organized people have the ability to retain and manipulate information or ideas. Like a computer-generated image suspended in space or a hologram in a sci-fi movie, information is "held up" to scrutiny, slowly turned around and considered from different perspectives, almost as if it were a three-dimensional object. You can consider representational thinking to be reflective—not gut-reacting, seat-of-the-pants thinking, as valuable as that can be in certain cases. This is the mind that takes information, steps back, considers and reflects—often looking at things in new and different ways.

Some people are more comfortable molding visual, verbal or spatial information. Martha Stewart is probably far better at solving a problem of how to decorate a certain-sized room for a holiday party than, say, Albert Einstein might have been. And vice versa if the information that needed to be molded involved theoretical physics. But both illustrate the same principle. No matter how it's done, or in what context, the ability to "turn over" the information after the stimulus is gone and do something with it—this is a skill to know, embrace, develop.

5. Shift Sets: People with superior muscle flexibility can touch their toes, demonstrating what exercise physiologists call "range of motion." In football, quarterbacks come up to the line of scrimmage and observe how the opposing team is arrayed to stop them. In the seconds before the play begins, a quick-thinking quarterback will call what's known as an "audible"—a last-minute change in what he is about to do, based on the quarterback's instant reading of the way in which the defensive team is positioned against him. This athlete's brain flexibility has equal importance to his physical flexibility.

The organized brain is ever ready for the change in the defense; the new game in town; the news flash; the timely opportunity or last-minute change in plans. You need to be focused but also able to process and weigh the relative importance of competing stimuli and to be flexible, nimble and ready to move from one task to another or from one thought to another.

In other words, you need mental range of motion and the ability to call an "audible" at your own "line of scrimmage." Because this is the way life presents itself, isn't it? To illustrate this cognitive flexibility and adaptability—the ability to shift sets—again consider the particular deficits of persons with ADHD. While those with ADHD are often considered to have a *deficit* in attention (as if he or she can't pay attention at all), the better description is that they cannot *regulate* attention. The mental switch is set to "on" or "off," and it's hard for them to change it back; sometimes they can't pay attention, but sometimes they can't *stop* paying attention, even when more important or salient stimuli are at hand.

6. Connect the Dots: The organized and efficient individual pulls together the things we've already talked about—the ability to quiet the inner frenzy, to develop consistent and sustained focus, to develop cognitive control, to mold mental/virtual information and to flexibly adapt to new stimuli. The organized and efficient individual synthesizes these qualities, much as the various parts of the brain are often brought together to perform tasks or help solve problems, and brings these abilities to bear on the problem or situation at hand.

The disorganized, unfocused individual may do none of this. We all know people whose lives seem to be out of control—and at the moment, you may feel like you're one of them. At times like these, it seems as if nothing ever gets done. You feel as if you're in a losing race with the clock and the calendar. You seem to have no ability to influence or manage events and "things just keep happening" to you. It seems as if there is no time to accomplish the important things.

You see where we're going here, right? Connect the dots: Thinking…feeling…acting…living. Following a logical path, from emotional control through the different cognitive building blocks, you are ready to put it all together. Here, the organized brain orchestrates all the other functions. The end result: a cognitive harmony that allows you to function more effectively, productively and enjoyably in every aspect of life.

One last time, let's go back to our example of Jill and her keys. In suggesting the idea of the launch pad to this patient, I was actually addressing two of the Rules of Order.

First, because she was emotionally distraught over what her episode with the keys had wrought in her workday, I knew that I had to calm Jill down; to help her Tame the Frenzy (Rule #1). You can't get organized and can't make rational decisions about *how* to get organized when you're distraught. In her case, the suggestion of the launch pad began a new process of thinking, not only reacting to the problem at hand.

Next, finding the little box that she eventually used for her launch pad and clearing out the space for it at home and in the office helped her to Sustain Attention (Rule #2) on the tasks at hand:

1) putting her keys down and later
2) finding her keys—by removing physical/cognitive distractions

This small success helped Jill become more confident. You can imagine her now starting her morning on a more positive note, heading out the door on time and ready to face the day, as opposed to already demoralized, frustrated and down on herself because of a moment's inattentiveness.

In the pages ahead we will examine more closely each of the Rules of Order, one at a time, and give you the tools and solutions that can help you to better sustain attention, stay on task and, above all, create a greater sense of order and efficiency in a world that often seems anything but.

Coach Meg and I will provide you with your own launch pad—and then some.

CHAPTER 2

A Change Will Do You Good/
Coach Meg

M Y COAUTHOR, DR. PAUL HAMMERNESS, does in this book what doctors do wonderfully well at their best—share their expert knowledge and wisdom in a compelling fashion so that you can make the best possible decisions and choices about your health and life. But of course knowledge and insight are only a start. Knowing what to do is one thing; knowing *how* you're going to do it is quite another. Doing something means that you need to make some changes, develop some new habits and unlearn some old ones. That's where I come in.

As a professional coach, change is my business. My kind of coaching has a few things in common with those who coach football or basketball teams. Like the men and women who exemplify the best of that profession, we know how to help people accomplish their best. But our goals are not to win games and the people we coach are not always young or athletic. Today the most established domains for coaching are in the executive suites of some of America's largest corporations. In the United States, more than five thousand executive coaches help CEOs and other leaders to improve their performance, impact and capacities and to handle the pressure cooker of the executive suite without sacrificing their families and health.

Of course, executive coaches are not limited to helping executives. Indeed, many of the people I've worked with were individuals trying to either get *to* the top of their field or *on* top of some other aspect of their lives. Coaches help clients navigate life transitions or realize lifelong goals and dreams. Wellness coaches work with people to improve their health and well-being in a way that lasts.

No matter whether we're helping Fortune 500 executives balance their lives or middle-aged sedentary individuals manage their health, the process and the tools we coaches use are often similar. In a nutshell, we help people organize their brains for change. I help you achieve clarity, choose a focus, build a plan, and embark upon and complete the journey of change. Doing this requires an understanding of how the brain works, and because of that, the best coaches have strong foundations in psychology and neuroscience. In particular, we are interested in the psychology of change. Research suggests that this work—the work of change—involves the activation and organization of the prefrontal cortex of the brain or, as Harvard psychiatrist and author John Ratey calls it, the "thinking CEO" region. Coaching also works on the limbic system, the home of our emotions. Coaches help clients increase positive emotions and better manage or decrease negative emotions, which increases the likelihood of success. Positive psychology research has proven that positivity, or increased positive emotions, opens and broadens our thinking and increases our resilience and capacity to change.

As a coach, my specialty is inspiring and facilitating the process of change—and that's what I intend to do with you in the pages of this book.

I am here to help you make the changes that will enable you to get a better handle on your life—to get your life better organized and to help you become more attentive, focused, and less distracted. My job is to prepare and guide you through this journey. I will help you motivate yourself (the only kind of motivation that works). I will help you identify and mobilize the resources you need. I will try to build your confidence so that you can complete this journey. I will keep watch for the obstacles and hazards along your journey—the kinds of things that can derail the change journey—and help you steer clear of them or get back on your feet when you fall back.

Of course, while I can show you the way, I can't take the journey for you. Making change is work; it takes time and commitment. The fact that you are reading this book suggests that you have already taken an important first step on that journey.

In the previous chapter, Dr. Hammerness identified and explained the Rules of Order, the traits demonstrated by individuals who are functioning at a high level of organization and productivity. So there you have the "what to do"—what you need to do in order to become the person who is on top of things in your life. Now comes the "how to do it" part.

As I've said, getting better organized and more focused is a process of change. And like any change in your behavior, whether it's losing weight or quitting smoking, it's going to require a certain mind-set. In order to achieve that mind-set, and to begin to get a better handle on your life and the changes needed to feel less frenzied and more in control, I'd like to share some tips with you, based on my knowledge and experience as a coach.

YOU'RE THE BOSS

Enough for a moment about Dr. Hammerness and Coach Meg and about Rules of Order and action plans. Let's put the spotlight on *you*.

As discussed earlier, humans are wired to want to be in control and resist being changed by others. It manifests itself early—witness an infant's need to assert even a modicum of control by refusing to eat his mushy carrots. It reaches another peak when a parent reaches her elder years and heartily resists the advice of her children.

It's up to you to decide that you want to have a better organized life—and that, in doing so, you need to some degree to change the way you use your brain. Maybe your spouse bought you this book in the hopes that you won't lose another pair of expensive sunglasses or forget that you left the pot of water boiling on the stove. That's fine; it is still up to you to decide to change—and on your terms. Forget what others are telling you that you need to work on. You're in charge. You choose.

We have spotlighted some of the key principles of organization and focus (our Rules of Order) to help give you the language, a reference point, a starting point. Ultimately, though, the ability to incorporate, to some degree or another, all six principles will go a long way toward improving the organization of your life, but that doesn't mean you have to master all six. Again, you're the one best suited to know.

GET CLEAR ON YOUR PRIORITIES

You can't change many things at the same time; you're not likely to change any if you're using up all of your bandwidth at this moment. Unlike the cable modem that can seemingly accommodate unlimited data, pictures and text, your personal capacity to handle data and stimuli has its boundaries. You may know that you need to get better organized at some point, but your first priority may be to help a colleague through a crisis or care for a sick family member. Know that you can put the book down and come back to it later—you may have more important priorities right now.

But let's say that getting better organized feels like the top priority right now. You're sick and tired of feeling distracted and disorganized during the day and worrying that not much got done by the end of the day. You feel it's time to get beyond the struggle that has plagued you for years.

Now you have some choices. Dr. Hammerness has presented his Rules of Order—those six areas where you can work on improving the organization of your brain. Reflect on and assess your mastery or lack of talent or skill for each of the six dimensions of an organized brain. Celebrate and be grateful for the dimensions where you are in good shape, either because you inherited good genes or you had an amazing parent or teacher who patiently and relentlessly helped you build that dimension. Appreciate that you can enlist the dimensions where you are strong to help you improve on the weaker ones.

Choose what, when and how to change carefully and thoughtfully. Success begets success. Failure will damage your confidence and bring negativity.

Although several areas may be calling for attention, it's important to pick the area that will set the dominos in motion, the area where you can make good progress quickly and build confidence in working on other tougher areas.

Which area might set in motion a domino effect (like unblocking a blockage)? Which would make your life better and open up new possibilities? Which one of these principles sounds like something doable to you? (When facing a number of changes, it's often a good idea to build your confidence by tackling the one that seems the one you can do with the highest probability of success.) Which area are you drawn to? Which feels like a good pain that is beckoning to be eased?

Rate each area's importance to you (out of 10) and your confidence in being successful (out of 10)—start with the area that has the highest score and, even better, a score of at least 6 on both ratings.

Let's say you've had a problem staying with one thing—a project at work or even a book you're trying to read. Then learning how to "sustain focus" might rate a high score. Or perhaps you purchased this book because your spouse has pointed out your inability to stop what you're doing and attend to something else without getting flustered. Maybe you've noticed that you get angry and frustrated and curse or bang on the computer keyboard whenever unwanted e-mails pop up, distracting you from what you're doing, and you're not sure whether to jump off what you're doing and address the e-mails or finish the task at hand. A smashed keyboard would certainly suggest a 10!

IGNITE YOUR MOTIVATION—THE JET FUEL FOR THE CHANGE JOURNEY

Your motivation is the jet fuel for the journey of change—the hotter it burns, the more likely it will get you through unavoidable and unpredictable setbacks, moments of doubt and any other stuff that pops up to throw you off course. So let's find something really flammable! Is there something in your life that's important to you that's being affected by your relative inability to focus the way you want to, your distraction or your sense of being overwhelmed by all the stimuli and messages competing for your attention?

Spend a little time digging down to the biggest benefit of getting more organized. Not one imposed by anyone else (like hanging onto your expensive glasses) but something that you get fired up about. Remember some of the statistics we presented earlier about the problems associated with distraction and disorganization: this could very well be causing sufficient stress that it might be affecting your health. Your job performance could be suffering. Or you could feel that you're spinning your wheels, not getting ahead in school, career or life. It could also be affecting your family life.

The best motivator is to connect the change to a higher purpose (something that hits you in the gut or brings tears to your eyes); how it will help you do the things that make you thrive, realize your life's purpose or legacy, or make a difference in your world.

Expand your motivator into a vision statement, such as:

I will improve my relationship with my children if I am better able to tame the frenzy and focus mindfully on our conversations.

I will appreciate the good things in my life more fully if I'm not distracted by the stressful "noise" in my environment.

I will be more creative and have better judgment if I detach myself from the hubbub of daily life.

I will get more done and feel better about how much I accomplish if I am not diverted and distracted.

Your motivation is truly the jet fuel for the change journey, both in the early phase as you build new fledgling connections and paths in your brain and later to keep you on track with new habits. If your motivational tank is low on fuel, you're not likely to be successful.

MAKE SURE THAT THE PROS OUTWEIGH THE CONS

You may have one really compelling reason to change the behaviors that are contributing to your continual sense of disorganization, distraction and loss of course—or you may have several. But the reasons *not* to change, at least right now, may win out. Or while you may decide to push forward, you could quickly find yourself back on the fence with second thoughts, weighing whether "to do or not to do." Psychologists call this a *decisional balance.* If you find yourself in such a mind-set, list the reasons to change in one column and reasons not to change in another. Even better, find someone to orally do this exercise with you. If the reasons to change clearly win, then you are ready to move forward.

No judgment allowed—if you feel badly about the reasons for not making the change and staying the same, let it go. You can't easily move forward with a rain cloud from the past following you around, even on sunny days. From time to time we all face deeper issues that hold us back, undigested life issues or unhealed pain. You need to heal old wounds or get a new perspective on life issues with a therapist or other program designed for that purpose.

BUILD CONFIDENCE TO MEET CHALLENGES

There will always be reasons to do nothing and to talk yourself out of making changes and meeting challenges. Getting your life better organized sounds like a lot of work. It's not the right time, you're too busy, it's football season, it's your son's graduation or your wedding or whatever. While there are periods in your life that may not be best suited for making major changes, doubting yourself as to the timing, whether justified or just a convenient excuse, will eat away at your confidence in your ability to change. Ask yourself on a scale of 1–10: "How confident am I that I will be successful in overcoming my challenge and making this change?" If your score is below a 7 then you should first spend a little time thinking through ways to handle your challenges. As Henry Ford said, "Whether you think you can or think you can't, you're right."

Make sure that you think you can.

Sometimes it's as simple as scaling back the goal a little so that it moves from "I'm really anxious about whether I can really do this" to "Absolutely!" Sometimes you need to shorten your horizon: take it one day, even one hour at a time.

Or you may discover that you need to learn a new skill and gain knowledge first because you're about to do something you've never done.

Set small first steps, and don't worry about how long it takes to make them. The race to long-lasting change is usually won by those who take time to build the foundation needed for new habits to last.

ADOPT THE MIND-SET OF A SCIENTIST

As much as we would all love a quick fix or shortcut and avoid a lot of experimentation, there is no one else quite like you. Someone else's prescription probably won't get you to the finish line. That's why our approach in this book is highly individualized. Here, you get options, you get choices and you get to pick what works best for you. Sure, we have some evidence-based principles to offer you; yes, I'm going to give you the information, techniques and approaches that I know can work, but just how they will work best for you, to what degree and how you'll integrate them into your day-to-day life…well, that's going to be up to you.

We'll talk quite a bit about the science of the brain and how it can help you. So get into the mind-set of a scientist. Be ready to do a few experiments, observe the outcomes carefully, think back to past experiences that might be revealing and decide which habits fit you best now, based on the results. Don't worry, I'll help. I'm here giving you some suggestions and guidelines and a framework to help you make these changes, but ultimately it's your experiment, and you're the one best suited to judge the results.

CALL IN YOUR STRENGTHS

Another way to cultivate confidence is to bring your strengths and talents to the table. It's very easy to forget what you're good at when you're swimming upstream. While your desk may be a mess, your kitchen pantry may be well organized. While you may feel unfocused and unable to stay on top of things, you may be quite capable of helping your colleagues organize *their* projects. Or you may be known as tenacious, as someone who doesn't give up, who's determined to close the deal, find the item you're looking for and reach the finish line of the race. Or you're creative and you have a knack for finding new ways to do things.

By the time we are adults, only one-third of us have a pretty clear idea of where our strengths and talents lie. We are typically much better at naming our deficiencies. If you want to learn more about your strong points, complete the Clifton StrengthsFinder assessment (www.strengthsfinder.com) or do the VIA (Values in Action) Survey of Character at www.viacharacter.org. You can also ask your family members or colleagues what they think are your strengths.

The important thing to understand is that whatever realm they lie in, with a little digging you will find that you *do* have strengths, talents and abilities. And those strengths can be used to improve or overcome those areas in life where you are not as strong.

FOSTER POSITIVITY

Barbara Fredrickson, an author and inspirational leader in the emerging field of positive psychology, has taught us that you need to be at or above the tipping-point ratio of at least 3:1 of positive emotions to negative emotions for your brain to function at its best. In other words, you need a 75/25 percent positive energy ratio to succeed.

This isn't just facile "put on a happy face" stuff. It's hard to be positive all the time. Some days, it may be hard to feel positive at all. But while negative emotions are good teachers, you can't change if your thinking and energy are impaired by too many negative emotions. Fredrickson has also taught us that positive emotions are the active ingredient enabling "resilience." This is that wonderful quality we so admire in children. It's caused by responding positively to adversity and is necessary for change. It's inevitable that you will fall back from time to time. In fact, if you don't bump into setbacks you're probably not going to make lasting change. But try to see these challenges as teachers and friends: welcome them and appreciate them and they will serve you well.

Here are some ways to "reframe" your emotions and accentuate the positive. One way is to make peace with the past. Negativity in one area (for example, not forgiving yourself or someone else for something that happened) can follow you around like a dark cloud that overshadows the otherwise positive aspects of your life and disposition. It's particularly important to let go of the past as it relates to the area you're working on. If you feel ashamed or embarrassed about your past behavior and performance, the negative feelings will act like brakes on your forward motion. If it was an embarrassing or damaging incident that prompted you to pick up this book—whether losing your keys again or losing a job because distractions got in the way of performance—well, you need to put it behind you, starting right now. The past is the past. What's done is done. Time now to take the lessons of what happened; apply them to what will happen and develop a fresh, open and positive outlook toward the future. Here's a little pep talk I give to my clients who are stuck on mistakes they've made in the past. If need be, you can use this "mistake mantra" to absolve yourself:

> I forgive myself for the mistakes I made. I'm not perfect—no one is— and I'm committed to learning and getting better. In fact, the past experience is my wise teacher, and I will apply the lessons well.

Enough with the mistakes. Fredrickson has identified the most common positive emotions. Here are some you can work on:

- Cultivate curiosity about and interest in the challenge of change.

- Seek inspiration from others who have been successful.

- Be grateful for something, anything.

- Savor small moments on the journey.

- Enjoy the pride of doing something well—appreciate even small steps forward.

- Celebrate early wins. It's very easy to ruminate on the negative. It's less familiar to focus on the positive.

- Have fun. Making positive changes in your life can be extremely enjoyable. We don't mean to make this sound like a lark; it's not and your reasons for wanting to get yourself "together" may be serious. That doesn't mean you can't discover joy in the process of changing. In fact, you probably will!

BUILD A SUPPORT TEAM

It's hard to change when your environment is working against you. A chaotic or noisy desk or office can be highly distracting. Or your spouse may be raining negativity on your time together. What can you do?

- Engineer support: Clean your desk or office. Ask your spouse to suspend the critique and say affirmative things for the next ninety days (after that it may become a habit!).

- Tell a friend or two that you're working on making some changes and ask them for support, via regular phone check-ins or e-mail reminders. Find a buddy with similar issues and work on your vision and goals together and meet regularly for mutual support.

- Celebrate progress together for more reinforcement. That could be with a spouse or a close friend. Ask your children to help and encourage you, perhaps by cleaning common areas of the house so you're not distracted. It's so much easier to change when you've got a team cheering you on!

CREATE A VISION FOR CHANGE

Creating a clear vision of your ideal destination is an important early and ongoing step for your journey. Neuroscientists at the Massachusetts Institute of Technology have shown that when people reflect frequently on what their positive future selves will look like, they are more likely to make choices in their long-term interest rather than shortsighted ones.

Who do you want to be? What do you want your life to be like? What's the best thing that will happen when you're more organized? Let's answer those questions.

The first step is to accurately understand where you are now. Self-awareness is a necessary precondition for change, and so your first step is to explore what's working and what's not working when it comes to the state of your organization at home and work. Reflect on where you are right now, and look at it honestly. Get real but don't beat yourself up. Appreciate that the past is your friend and focus on how it helps you go from here.

To help you figure out where you are, here are a few questions that allow you to put a number to something that isn't easy to quantify—where you are now and where you want to end.

Self-Assessment Questions	Today's Rating	My Target
What percentage of my workday feels disorganized or chaotic?		
What percentage of my home life feels disorganized or chaotic?		
Rate the level of disorganization in my worklife out of 10 (10 being highly organized).		
Rate the level of disorganization in my home life out of 10.		
What are the top three things making me disorganized?		

1. _____

2. _____

3. _____

Now you have a better sense of where you are, where you want to get to (your target) and how far that is from the current picture. From Dr. Hammerness's chapter, you've also begun to understand some of the factors behind organization (and lack thereof), as well as the things you need to aspire to in order to change that. Now it's time to create a vision, one that inspires you and one that makes you feel hopeful and optimistic. As your coach, I want to help walk you through this "vision creation" process, which is specific and clear and can be enormously effective. We do it through a series of questions that we call a Vision Grid.

Vision Grid	
Vision Statement	A vision is a compelling statement of my ideal state when I'm better organized. What would I look, feel and act like at my ideal? *I would feel a lot more in control than I do now, like I'm on top of things and I'm doing well.*
Values & Motivators	What element(s) do I value most about this vision? Why is it really important to me to reach this vision? What good will come from my doing so? *I want to contribute fully in my life, on the job and at home, and I can't do that now because I can't seem to get things together.* *I sometimes feel like my relationships and my employment are in jeopardy because of my inability to stay on top of things.*
Gap	How large is the gap between where I am today and my vision? How long will it take?

Gap	Right now, at only about 30–40 percent of the time do I feel like I'm on top of things in my life. I want to get it to 70–80 percent.
Strengths	What strengths can I draw upon to help me realize my vision? How can the lessons from my life's successes be applied to reach my vision? *I'm willing to make a change and I'm eager to learn; I'm the sort of person that likes to learn and improve.*
Challenges	What significant challenges do I anticipate having to deal with on the way to reaching my vision? What concerns me most? (Note: Here's a very common concern among people who are having difficulties getting organized in their lives.) *I'm good at getting started, but I have trouble following through and completing things.*
Strategies	What strategies may be effective to help me rise above my challenges? (Here the choice is yours: Which do you think has the highest likelihood of your success? For this example, let's list three strategies to address the challenge of follow-through.) • *I've got a project deadline eight weeks from now. I will break it down into discrete steps or stages, thinking about where I want to be with this project in one week, two weeks. So instead of worrying about what I'll be doing a month from now, I'll think about what I'll accomplish week by week. This way, I'll set up a series of intermediate goals, and won't lose my pace or focus.*

Strategies	• Lack of follow-through is often a sign of my flagging enthusiasm. How do I recharge my motivational batteries? I'll put a Post-it note on my desk saying, "I'm on top of things" and place a picture of my family next to it, reminding me of the importance of addressing this issue. • Generally, some aspects of any given project are more interesting than others. Tedium can often lead to failure to complete or follow-through, so I'll break down the big project and look at the more interesting parts. Again, I'll write them down to help remind myself of these more rewarding and challenging aspects of the project. I'll make a decision to focus on the more mundane parts today, but "reward" myself by looking at the interesting "big picture" things tomorrow. This will help keep my eye on the more fun and creative aspects of the project and, again, will help keep me on top of it.
Supports	What people, resources, systems and environments can I draw upon to help me realize my vision and meet my challenges? I've got a friend at work or a tennis partner or my spouse that I can talk to about work. I'll make sure I discuss the progress of this project a couple of times over the next eight weeks. Talking to him or her will help keep my eye on the big picture and will enable me to have a broader perspective on the progress of the job. I want to make sure I get my exercise in that week and get enough sleep so that I feel fresh and energized and ready to work on this project every day.

Supports	*When I want to do some of the big-picture, creative planning on this project, I'm going to have some background noise. I'll go to a coffee shop and put on some classical music.* (Contrary to what you may have heard, having background noise can help tune out outside distractions and help focus.) Oh, and keeping reading this book!
Confidence	On a scale of 0–10, with 10 being really confident and 0 being not confident, how confident am I that I can close the gap and realize my vision? *I feel energized because I have a vision and I know why it matters. I have some strategies and a support network in place, so I'd say it's now a 7 out of 10. Let's go for it!*
Ready & Committed	Am I ready and committed to take the first steps toward my vision? (We'd like to answer that for you, if we may.) You've already purchased this book and read this far. That's a very good sign that you're ready to commit. Let's shake on it!
First Steps	What first steps do I want to take? (Big journeys of change often start with small, seemingly innocuous steps. They could be very practical or symbolic—different people respond differently.) *Again, using the example of "I want to stay on top of things," I will:* • *Put a photo of me on top of a mountain* (or, if you've never been to the top, a picture of Sir Edmund Hillary on Mount Everest or any other wonderful, inspiring summit shot you can find

First Steps	in *National Geographic* or online. Don't scoff at the power of small "totems" to motivate you big time).

*   *Make a list of the first five small steps I'm going to take on this project.*

*   *Reorganize my workspace. Not necessarily a complete, ergonomic redesign. Again, it will be just a small adjustment: I will finally move that stack of reports that have been sitting on my desk for weeks. I'll clean off my computer screen. I'll reposition my chair, so now I'm looking at the picture of the mountain I've hung up!*

(This is less a functional change as it is a way for you to take a fresh look at things and to help rededicate this workspace as a place of action, confidence; a place where you are now going to start being on top of things!)

In the next chapters, after Dr. Hammerness explains further each of the six Rules of Order, I'll show you how you can specifically adopt these principles of organization as your own. You can also adapt this sample vision grid and tailor it for your own use—or use it as a template to create a vision of how you're going to conquer the organizational challenges in your life.

CHAPTER 3

Rules of Order/*Tame the Frenzy*

C ONTROL YOUR EMOTIONS OR THEY WILL CONTROL YOU"
are the words of a famous Chinese proverb. We begin our Rules
of Order—our first step toward getting our lives better organized and
more in control—by working to better manage our emotions.

We are human because of our capacity to feel and to experience
an emotional life. But, right from the beginning, emotions can block
the entrance to our path to becoming better organized. Emotion and
cognition—feeling and thinking—must be integrated in order for us to
function at our best. In this chapter, we discuss the remarkable neuroscience of emotional control and how achieving a balance of feeling and
thinking is a fundamental prerequisite for the organized brain.

Emotions are as varied as we are: the so-called "primary" emotions—
anxiety, sadness and anger—are, like the primary colors, basic and invio-
lable. But just as on the artist's palette primary colors can be combined
into dazzling new creations, so, too, can our emotions meld together
to form every hue imaginable.

Because of the great range of feelings, some theorists have organized emotions around two set dimensions, each encompassing a broad range of feelings: valence (pleasant to unpleasant) and arousal (calm to excited).

Others, like the famous psychologist and author Richard S. Lazarus, separated human emotions into several distinct categories. There were the "nasty" emotions (anger, envy, jealousy), the "empathic" emotions (gratitude, compassion), "existential" (anxiety-fright, guilt, shame), and those provoked by life conditions, both favorable (happiness, pride, love) and unfavorable (relief, hope, sadness, depression).

No matter how you categorize them, emotions take many names and come in many forms: worry, panic, tension, stress, sadness, despair, frustration, irritation, exasperation. Emotions can be felt, and emotions can be voiced. Emotions can take the form of a sudden surge of anxiety or a frustration that grips us in a moment. Emotions can be experienced as quiet ruminations others don't see. Sadness, anxiety, anger—the blue, yellow and red of our emotional color scheme—can be interrelated, can feed off each other and can take turns in disrupting your organized plans, your organized brain. To extend the artistic metaphor, these emotions can take the orderly hues and shapes of your life and make them look like a Jackson Pollock canvas.

For those who feel disorganized, overwhelmed or caught up in a frenzy, those "basic colors" of the emotional palette are typically their root emotions. While we're all familiar with them—perhaps all too familiar—I thought it might also be useful to include their definitions:

- Anxiety (worry or unease about what may happen)
- Sadness (a state of unhappiness, sorrow)
- Anger (irritability, hostility)

How do these common emotions manifest themselves in the situations that we're dealing with in this book? You can feel anxious about the implications of your disorganization ("What is losing that important document going to mean when I get into work next week?"); sadness about the impact on your apparent inability to change ("Why can't I stop losing things?"); anger about the challenges at hand ("I'm going to have go back and redo hours and hours of work because I was so stupid!"). I've had patients who have exhibited all three and who, in their respective quests to get a better handle on their lives, have had to wrestle with these basic emotions. Anxiety, sadness, anger.

Let's meet one of them.

CASE STUDY IN ANXIETY: THE ONLY THING WE HAVE TO FEAR IS FEAR ITSELF (AND MAYBE XBOX, TOO).

The woman in her late thirties who walked into my office—fifteen minutes late for her appointment—was clearly distraught. Her eyes were red, she looked as if she had been crying and her face was careworn. She appeared as if she hadn't been getting enough sleep. Her primary care physician had referred her to me, apparently with good reason.

"So," I asked, as she sat in the chair in my office, eyeing her surroundings warily, "did you find the office without any problems?"

"Not really," she said, rolling her eyes. "I'm sorry I'm late. I made a wrong turn in Central Square and practically ended up in Somerville." She laughed ruefully. "It's just another example of the shambles my life is in these days."

I waited a beat, but she said nothing further. I tried to gently prod her.

"Can you elaborate on that?" I said. "How so?"

She sighed deeply and then related a tale of unhappy events. Eileen, as we'll call her, had been divorced a year ago. Her son, the product of that marriage, was twelve years old—a "tween" as the demographers now call these early middle-school-aged kids because they are between their childhood and teenage years.

The son, to hear her side of it, was struggling: in his first year of middle school, he had a heavier workload, and a bulging backpack to go with it. He had band practice, baseball practice, tests to study for, homework to do, friends he wanted to hang out with and video games he wanted to play. It sounded, to me, like a fairly typical schedule for a sixth grader these days— school, music, the arts, sports—but instead she told me they were one step away from everything falling apart, it seemed, almost every night.

"The other night, he was late to his baseball game," Eileen said. "It was because I was behind on some of my reports at the office, and I got home a few minutes late from work. But then I couldn't drag him off the Xbox game. This had happened a couple of times already, so the coach said he couldn't start that night, and he had to sit on the bench most of the game."

I started to interject, but she was already off on the next of what became a litany of catastrophes: There were problems keeping up with his schoolwork. There were problems at her work (she was a physical therapist). There were problems with other members of her family and with her ex in-laws (or as she preferred to call them, "the outlaws"). She wasn't whining; she just sounded weary and overwhelmed.

She went on to tell me how she had responded to some of these crises: she talked to her son's teachers. She got a tutor to help him with math, which seemed to be his most challenging subject and was not her strong suit.

I had a sense that she and her son often worked things out reasonably well, but she still seemed to be operating in constant crisis mode. "What does your son say about all this?" I asked.

She rolled her eyes again. "He says, 'Mom, stop stressing out.' Easy for him to say."

I had a feeling her son is right. I was hearing disorganization here, yes, but I was also hearing anxiety and worry, and I was glad that her son had articulated it. I jumped on that.

"Why do you think he said that?"

She pondered the question a couple of seconds. "I guess I overreacted to getting to the game late," she said. "I told him that because of his video game he'd be thrown off the team, and so I told him no Xbox for a month. But I also called the coach and told him it was totally unfair to bench my son anyway. He's been playing really well, and so what if he was late a few minutes a couple of times. Who isn't? I mean the traffic alone...."

Then, striking a mildly defensive tone, she sat up straight in her chair and added, "And in fairness to me, I *do* have a lot to be stressed about. And isn't that normal in this day and age?"

"Well, it's not about what's 'normal'; it's about what's right for you," I said. "And as you've been talking about how disorganized your life is, it does sound as if your stress level has been equally high."

She nodded her head in agreement. (This was important—I was glad to see her acknowledge the emotion.)

I continued, "So you're feeling disorganized and you've also been really stressed and anxious. One of the important questions at a given moment or during a given day is what comes first... the disorganization or the anxiety?"

She raised her eyebrows as she pondered that. I went on.

"They can fuel each other. You might see that one precedes the other or one makes the other much worse. For example, if you hadn't felt so...what was the word—*panicked*?...about him continuing to play the video games even after you'd arrived home late...or to his being benched...you probably would have found a better approach in dealing with both him and the coach."

She nodded. "I guess so...maybe."

"What I'd like you to start doing is keeping track of this balance or imbalance of stress and disorganization," I said. "Take a reading of your stress, day to day. Think about when you're stressed, how often it happens, what you are feeling and what you are thinking at the time, and see if you can identify patterns. The first thing we might need to do is get a handle on that stress and anxiety. I suspect that will give us a better starting place to work on organization."

This was just a small start. As with any patient, all of this would require work; there were other issues, and I don't want to imply that one appointment made all the problems in her life go away. However, this was an important first step in tackling her sense of feeling overwhelmed and disorganized and that her life was in "shambles." Eileen's reaction to these situations was part of the problem. It wasn't that she got "stressed out" because everything was in total shambles; she got stressed out first and, very often, disorganization followed and increased from there.

It didn't have to be that way; getting a handle on her emotions was the first step —and I'm happy to say, she has made great progress in doing exactly that.

THE FEELING BRAIN . . . THE THINKING BRAIN

Sometimes it may seem, as it did to Eileen, that we are totally ruled by our emotions. Not true. Your brain has spent a lifetime evolving and has the inherent capacity to handle your frenzy. It can allow the emotions to enrich our lives and not wreak havoc on them. Remember that the human brain has developed from a very rudimentary organ, with primitive reactive abilities, to one that is stunning in its complexity and in its ability to think and to feel and to manage both. Remember, also, that the primitive human brain wasn't designed to deal with a barrage of tweets and text messages. Its first job was to keep you alert and alive. Over millions of years, that basic, primitive brain has been built upon with layers of complexity—from the primitive brain to the emotional brain to the rational or intellectual brain, but one that can still feel, love, imagine and dream. Indeed, the brain may be continuing to evolve right now. Perhaps it will eventually adapt to the increasing levels of stimuli around us and form additional ways to better process it (which means that instead of reading this book and trying to better organize yourself now, you could wait a millennium for the human brain to evolve into some sort of "post-distraction" design, where it can process digital tweet-like squirts as easily as the modern brain can process language and writing).

You could think of the brain as a committee of experts, all the parts working together but each with its own special properties. As is the case with most committees, there is a chairperson and a hierarchy; each of the members is assigned their respective duties. In your brain's "committee," at the bottom of that hierarchy, are its primitive areas (also known ignominiously as the hindbrain), such as the brainstem. Here, basic vital functions occur, including the support of breathing and heart rate. If you ever wanted to get in touch with your inner caveman, here he is: the hindbrain supports our animal instincts, involved in simple "flight or fight" primitive reactions to the environment. When the heart rate is going, blood is pumping to our legs and providing energy to run away as fast as those legs will carry you. (This may help explain why when we are overwhelmed at work our first instinct is often "I've got to get outta here!")

The more sophisticated aspects of the brain evolved over and around these basic areas. The middle area, developed over the brainstem, is a complex area—a key emotional and information center. Areas within this region include the hypothalamus, thalamus, hippocampus, cingulate cortex and basal ganglia. In the 1930s, James Papez, an American neuroanatomist, proposed the idea that emotions were created through a circuit, connecting various parts of the brain. The Papez circuit—or limbic system, as it has come to be known—envisions a "mechanism of emotion" to describe the way in which our feelings are created and the subsequent pathways of communication and activity: emotional processing, emotional memory, and complex hormonal and motor reactions.

The amygdala (pronounced *a-MIG-da-la*) is a critical area in this emotional circuit. Long described as central to fear conditioning and response, as well as processing of positive information like reward, the amygdala evaluates information that is relevant to us and is involved in directing our response. Unlike our prehistoric ancestors, we no longer rely on the amygdala to help assess whether a saber-toothed tiger is lurking in the woods outside our cave. But we can benefit from a brain that is still vigilant and alert to the more sophisticated, subtle threats—and opportunities—in our modern surroundings. For example, Dartmouth College's Whalen Lab, where some of the country's leading-edge research on the amygdala is done, has found that facial expressions produce "robust activation" of the amygdala—meaning that this part of the brain is seen as active and switched on based on blood flow to the area or its use of the brain-food glucose. The scientists theorize that frowns, sneers, smiles or furrowed brows are "conditioned stimuli"; that is, such facial expressions have predicted certain outcomes in the past (the brain remembers the smiling face of the person who handed over your bonus check; that the nervous-looking visage, unable to make eye contact, announced the downsizing at your job).

While the brain areas and circuits that produce emotions may be as well-ordered as a municipal power grid, we all know that emotions can be messy. Modern-day neuroscience supports what is intuitive—that these emotional areas of the brain can interfere with even simple cognitive tasks. For example, a 2010 study at the University of Waterloo in Ontario, Canada, found that people who feel anxious while doing math problems can have trouble completing a task as simple as counting past five. This experiment shows that emotion can interfere with rudimentary brain processes.

The direct analogy is clear. When "turned on," these emotional centers of the brain can interfere with the basic building blocks (such as attention and focus) of more complex organization. When you are reacting emotionally—whether you're anxious, sad or angry—you are not thinking well. Studies consistently show how emotion can grab your thoughtful attention and turn it away from the task at hand. So it stands to reason that when you are trying to get to an organized foothold in what may seem like a chaotic life, the last thing you need are emotions that can sway, distract and eventually send you tumbling back down into disorder.

Ah, but the brain has an answer to emotional distractions in its higher, or "executive," functioning areas. To go back to our committee metaphor, the executive would be the chairperson. Another way to think of it is as the "central offices" of the brain—the thinking centers in the brain's cortex (those classic thick folds and crevices seen over the top of the brain) that direct the underlying older, "subcortical" regions (out of sight)—much as a supervisor's office would develop and direct the actions of the workers on the factory floor. The prefrontal cortex is one such critical cortical area, one that we will come back to time and again throughout this book. As we will see, cortical areas of the brain are involved in emotion and, more importantly, emotional *control*.

How do we know this? While scholars since Aristotle have theorized about the structure and operations of the brain, we can now describe that structure and the interactions of its various parts with some confidence because we've been able to *see* it in action. Brain imaging scans have progressed from being able to document the size and shape of the brain and its components to demonstrating the actual functioning of various regions in action. Some of the most fascinating functional brain scans involve special chemicals (radionuclides) that can travel in the blood to specific areas of the brain and light up for the imaging camera. These chemicals indicate brain activity during a particular task, such as looking at a picture, solving a math problem, or seeing a scary face. Functional brain imaging scans can offer incredible detail of brain activity. These brain scans have informed us about the nature of emotion and cognition and their delicate, dynamic balance. It's a real-time image of the human brain at work.

Okay, now let's go back to the amygdala and see how Eileen's behavior helps us reveal the inner workings of the "disorganized" brain—but also what the brain can do to get things back in order. We know that the amygdala is part of a fear network in the brain. This region can be identified in functional brain scans, as it "lights up" in the presence of anxiety or fear-provoking stimuli. (Some studies have even used exposure to snakes and spiders to elicit anxiety.)

And here's another important aspect of what these studies have found: When the amygdala is "acting up," the cortical, thinking areas of the brain seem to "quiet down." In other words, as we become more emotionally heated, we see a lessening of cognitive control, almost as if the emotional centers of the brain are "shouting down" the rational side. Quite literally, the hotheads have taken over!

THE BRAIN AND WILLFUL CONTROL

What happens when people try to change or control the fear, manage emotions and keep their thinking centers humming steady in the face of hot tempers? Can we cool off and calm down the hotheads? Can we master our frenzied emotions?

Functional brain imaging studies (commonly known as magnetic resonance imaging studies, or MRIs) have shown that when subjects use cognitive strategies to try and reduce negative emotions, an increase of activity can be observed in the cortical, thinking brain, such as the prefrontal cortex (PFC). These studies, done at the University of Colorado, show that the PFC can dampen the hot-and-bothered amygdala, just as a calm voice of reason can quiet down an over-agitated crowd or children in a classroom. In other words, when you consciously make an effort to think; to be rational; to get the PFC up and moving; it can respond and assert itself, telling the amygdala, in essence, to calm down.

What kind of cognitive techniques does your brain use to lasso those emotions and hold them down until they stop kicking and screaming? One widely recognized strategy is called *cognitive reappraisal.* Reappraisal, as the name suggests, involves reappraising a situation, taking a new viewpoint. This is a form of cognitive reinterpretation. It means that our brain creates a new meaning to a situation, and the result of that reinterpretation alters the impact.

During reappraisal strategizing—when one is considering a fresh viewpoint—reduced activity in the amygdala with increased prefrontal cortex activity can be observed. The implications: Your brain has a way of controlling your emotions and, in particular, your negative or counterproductive emotions. If we allow it to, the brain's voice of reason *will* be heard.

This is what Eileen did: Once she began to identify the patterns of stress in her life, she began to reassess or reappraise some of the predictable hot-button situations and think in more rational ways about how to respond. For example, she began to think that perhaps her son was dawdling on his Xbox not to deliberately irritate her. Instead, she thought, it may be that he was feeling a little anxious himself—about his talents as a ballplayer and his status on the middle-school team—and was using video games as a moment of escape or solace. She also began to see his actual provocations as evidence that he, too, was tired from his heavy workload and maybe a bit stressed out, too. "I thought all the time he was just doing this to get under my skin," she admitted during one session. "I guess I should have realized that a kid can get stressed out and tired, too."

You'll learn more about how to do this on a behavioral level later in this chapter, but first, remember those primary emotions—anxiety, sadness, anger—the ones most likely to be associated with those who feel disorganized, distracted and overwhelmed? We've explained how reappraisal can be used in a clinical setting to tamp down anxiety, as was the case with Eileen; now we will see how another one of the most common negative emotions can affect our sense of disorganization—and how we can use our understanding of the brain to deal with it.

Let's head back to my office.

CASE STUDY IN SADNESS: CHALLENGE THE MIND, CONTROL THE EMOTIONS

For someone in her mid-twenties, a vibrant time when life should be filled with promise and discovery, the young lady in my office seemed unusually sad and dejected.

Jennifer was a paralegal, working in a law office in the Boston area. She came to me because she was having increasing troubles on the job and was feeling disorganized and demoralized. In our first session, she expressed great concern over what had happened that day.

"I didn't get everything done yesterday that I should have," she said. "And when my boss called this meeting and turned to me and asked me to present some information, I wasn't prepared. We had to reschedule the meeting, this whole case is set back a week and it's all because of my stupidity." Her eyes moistened, and she took out a tissue. "I don't know how I got this job," she said. "Really. The partners are so smart...I don't know why they give me responsibility for some of this stuff."

As we talked more, it became clear that she wasn't just reproaching herself after one bad meeting. It turned out that the reason she wasn't prepared for the meeting was that the day before, instead of reading the material she needed to read, she had spent the majority of the time thinking about her own perceived ineptitude.

"Jennifer," I said, "it sounds like you're getting pretty down on yourself."

"Oh, I know," she said. "I do try to think positive. This morning, I knew there was a good chance I would mess up in the meeting, but I told myself to think positive and hope for the best."

That's a pretty fleeting positive thought. The "hope for the best" part didn't appear to be brimming with confidence over the prospect of a happy conclusion. "What did you do after things went wrong in the meeting?" I asked.

"Well, I apologized to my boss and then went right back to work. I figured the sooner I got back into the stuff I was supposed to have read, the better. That's on my agenda for tomorrow as well."

"Tomorrow's Friday," I said. "What are you planning to do this weekend, in terms of what's been going on at work?"

"I'm going to follow my mom's advice. She suggested that I just rent a movie and chill out by myself. It's been a draining week, and I think I'm going to hole up in my apartment and relax."

Uh-oh.

I heard some things here that I did like—Jennifer was trying to think positive and was showing some resilience; she was ready to "get back on the horse," so to speak, at work.

But while we usually hate to contradict a mother's advice, we had to in this case. Here's why.

IT'S NOT JUST THINKING; IT'S WHAT YOU'RE THINKING *ABOUT*

A recent study at the University of Wisconsin–Madison School of Medicine and Public Health showed that those who have a tendency to get stuck in sad thoughts might have a harder time getting out of them. In the study, subjects exposed to difficult images deliberately designed to elicit negative emotions (pictures of car crashes and so forth) were asked to envision a positive outcome. The researchers found, as with other studies we've mentioned, an increase in prefrontal cortex (PFC) activity, as the thinking parts of the brain sprang into action. However, some of the subjects in the study were clinically depressed. For these individuals, despite efforts at positive thinking, amygdala (the fear, threat sensor) activity remained high. By contrast, people without depression were able to reduce amygdala activity during positive thinking strategies and the harder they tried the greater the effect. "Those (healthy) individuals are getting a bigger payoff in terms of decreasing activation in these emotional centers," wrote the authors. By contrast, for those stuck in a depressed state, greater effort seemed to yield more amygdala activity, not less.

Such was the case with Jennifer. Whereas thinking rationally— "turning on" the thinking centers of the brain—helped to calm Eileen's anxiety, the exact opposite happened with Jennifer. At present, she was stuck in an emotional rut; more thinking just yielded more negative thoughts—and thus, she became more caught up in the negative emotions.

There's one more study to underscore this point. In a 2009 study, researchers in the Netherlands looked at how "keeping busy" works when it comes to emotional control. Instead of just envisioning positive outcomes in the face of stress (images of angry faces or injuries), these subjects were given what psychologists would call a "cognitive load"—in this case, math problems of increasing difficulty. Through brain imaging, the researchers found that the more the cortical areas of the brain were engaged by the math, the "cooler" the emotional areas became. The lesson here? There's thinking and then there's *thinking*. Going off to a quiet place to ruminate after an emotional event may not be the best approach because it allows for more sadness, more negative thoughts and more prolonged amygdala activity. Instead, we should actively engage the brain in a cognitive capacity—thinking positive, reappraising or maybe just any (nonnegative) thinking—and make those cortical areas get to work, thereby taking the stage away from the emotional centers.

Jennifer, I suggested, needed to make a concerted effort to do *something*. Put in a few extra hours at work, read a book or play a challenging game. Remember what we learned from the brain image studies: When the "thinking" parts of the brain are working, they tend to have a cooling effect on the heated, emotional parts. Plus, we've learned that the harder you work on your cognitive tasks, the greater that effect.

So, instead of heeding her mother's advice (sorry, Mom) and spending the weekend at home alone, potentially getting dragged deeper and deeper into her sadness (as the depressed subjects in the study did), Jennifer made a plan to get active—intellectually active—which she did, both in the short and long term. After a "pretty decent week-end," soon thereafter she went on to join a book club and volunteer to be on a fundraising committee at her church. All this activity and engagement replaced sitting around and thinking about how sad she was. The job kept her challenged, and in small steps, success led to success. Before long, she was performing up to par, making fewer mistakes and distracted less and less by her sad, negative emotions. Part of the reason she's not letting those emotions get the best of her is that her increased cognitive load has quieted those feelings, which in turn has given her time to rethink her situation (remember that "reappraisal" strategy?). After awhile, Jennifer realized she wasn't so "dumb" after all, a feeling that was reinforced when, thanks to some of her more focused efforts on work, she got praise for the quality of her work—and a raise!

Jennifer's frustrations resulted in sad ruminations. With other people, frustration and overload can provoke very different reactions.

CASE STUDY IN ANGER: LET IT GO!

I could tell that Mitch didn't really like the idea of coming to a psychiatrist. Mitch was in his late fifties, had put two kids through college and enjoyed a very successful career as a mortgage broker. Mitch had it all. He was a former high-school football player, I learned. Although the glory had dimmed, you had a sense that he was probably brimming with swagger and confidence at the peak of his career.

But then came the mortgage crisis, followed by the stock market crash and the recession. Mitch's company went out of business, and he found himself unemployed. To his credit, he had picked himself up and started his own consulting business, working out of his house in a very nice Boston suburb. But he was having a tough time. Business wasn't great—it wasn't great for anybody at that point—but he was in no danger of starving and could pay his bills. So what brought him in to see me, I wondered?

"My wife," he said. "She's worried about my anger."

"Are you angry?" I asked.

"Sure," he said. "I get ticked off sometimes."

This, I learned, was an understatement: Mitch's anger could be quite obvious. He was doing his share of cursing out loud, slamming phones and pounding on the desk of his newly appointed home office. But the other kind of anger, perhaps the more insidious type, had revealed itself after he had to do his taxes.

For the first time, he was filing as an independent business person. To save money, he didn't go to an accountant and tried to do it himself—just as he'd done all the years he was working for his previous company and filing one W-2 each year. Now, he had a drawer full of new and unfamiliar forms, some spotty records and lost papers. He'd always had an efficient secretary to keep the voluminous amounts of paperwork associated with his work. Now, he had to do it on his own, and it turned out he didn't do it well. Important documents had been lost, he ended up making a mistake on the tax form and it cost him time and money to rectify it. Mitch was angry. Very angry. But here's the key part of his anger that we learned only after subsequent sessions. While Mitch had been filling out his tax forms and working out of his home office, he had been seething with anger as he thought about the circumstances that had left his firm bankrupt. These angry ruminations were under the surface but still a major problem. He began to think about things that happened in the past: If it wasn't for his greedy old boss, the firm might have stayed solvent, and he wouldn't be home right now. He'd still have Marge, his trusty secretary, to take care of things. He'd still be free to concentrate on his strength, which was working with clients, not these damn details and paperwork. It wasn't just his old boss—it was the guy at the bank, former clients, the people at the IRS, two U.S. presidents, you name it. The more Mitch thought about it, the more people he found to get angry with and the more situations he could find to replay and rekindle his anger.

What's happening in Mitch's brain—and what are we going to do about it?

THE BRAIN IN ACTION: ANGER IN EVERYDAY LIFE

Anger and frustrations occur, and quite often the overt, in-your-face anger is what people key in on; that kind of anger is typically a brief, time-limited response. Very often, the anger that I see in my office arises in the context of an irritating, challenging situation, like when the person feels they are faced with a task that is not possible or reasonable. Or the anger may have arisen, as was the case with Mitch, because a mistake was made. Errors and poor decisions are often made by people who are disorganized at home or on the job. Because of this, we consider anger as one of the emotional "primary colors."

You realize that you lost an important document, forgot to return an important call or made a real gaffe because you were too distracted by other things. You're angry! So what is the brain doing at that moment—and even more importantly, what does it do afterwards? Are we aware of these angry ruminations?

In one recent study on the subject, researchers analyzed the responses of healthy college students to anger provocation. First, by asking some background questions, they found subjects who tended to act with displaced aggression (taking out their anger later) as opposed to showing anger in the moment. What they found in the subsequent brain scans was surprising. You could see patterns of brain activity showing persistent angry ruminations when provoked. Further, activity in the hippocampus—the memory area of brain—right after the anger provocation was associated with these subsequent ruminations. So while these students were not reacting with anger in the moment, their brains appeared to be dwelling on angry memories, like Mitch did.

So how can this below-the-surface anger be managed? A psychologist at Harvard University, Christine Hooker, has done interesting work in this area: she and her colleagues studied brain activity after arguments with partners, in order to assess brain activity in the days following a fight and, importantly, to link it to how we feel afterward. Those subjects with a greater level of PFC activity (remember, the PFC is part of the "thinking, rational" cortical brain) reported a greater ability to bounce back emotionally after a fight. These subjects demonstrated greater cognitive control in other tests in the lab as well.

Clearly, Mitch's wife was onto something when she steered him toward therapy. He *can* get "pretty ticked off," as he says. But counter to what he thinks or tries to believe, he is not managing his anger well. His hippocampus—his brain's memory center—is packed with vivid memories of people at work and situations he's angry about, some of them going back years. Because Mitch is busy being angry at his boss for the mistakes that led the firm into bankruptcy, he doesn't pay attention to the work at hand. And then because he makes a mistake, he gets mad—and dwells on that mistake for days and weeks to come, making him even angrier. It's a vicious cycle.

"These forms are going to kill me," he said during one of our sessions.

"Yes," I respond, "that really is possible." I was only half-joking. Clearly, Mitch needed to manage this anger, not only for the sake of getting his life better organized but for the sake of his health.

He nodded. "I know, it's not good. And I'm the first to admit that paperwork is not my strong point. But you see in the old firm, I had someone to do this stuff for me."

"Maybe you can do it here, too," I said. "Find someone who can help you part-time. A bookkeeper or someone who can come in and take care of the forms and paperwork."

That way, I reasoned, Mitch could play to his strengths. He could make client calls, help broker deals, be the idea guy—and, of course, generating ideas takes cognitive work. I envisioned Mitch's PFC working away, cooling off his emotional centers and putting some healthy distance from those angry memories.

That's what Mitch did. He got down to business—or at least the kind of business he enjoyed doing best. In letting go of the parts he was flubbing and instead concentrating on the other aspects of his job, he was more able to move on, keep busy and be successful. He was able to let go of the angry memories because he started to see them as unhealthy.

The funny part is that once he had tamped down his anger, he probably could have taken care of all the paperwork himself. In the calmer, less-distracted state he's in now, he wouldn't be as likely to make mistakes.

FRENZIED CONCLUSION

Frenzy happens.

Anxiety, sadness, anger happen. These emotions are part of the human emotional palette. But the good news is that they can be checked and handled. The science has now revealed to us the brain mechanisms that both feed and tame the frenzy and has also showed us that the more we work at it—the more we work to control our negative emotions— the more effective our efforts can be. They are efforts worth making.

Because when you have calmed your frenzy, you will have the opportunity to be better focused, less distracted and more organized.

Let's find out now exactly how we do that.

COACH MEG'S TIPS

Frenzy is an emotional state, where we feel a little or a lot out of control, making us agitated and edgy. Dr. Hammerness has beautifully described the three root emotions that underpin this state that we call "frenzy": anxiety, sadness, anger. The opposite of a frenzied state is feeling calm and peaceful, even when we're engaged in high-energy activities—and that's where we want to get you!

Unfortunately, for many of us, feeling frenzied isn't an occasional or fleeting state. It has become the prevailing weather, a cloud of emotional stress following us around for most of our waking hours. Even our dreams can feel frenzied. Some call this "hurry sickness"—we're always in a hurry, constantly rushing, fretting and rarely experiencing the opposite state of calm wellness.

This sickness is brought on by two sources: external, the frenzied world around us, and internal, the frenzy we create for ourselves. Some of this internal frenzy we are conscious of and can name and recognize. Some of it is subconscious, seeming to come from a source beyond reach.

External frenzy is everywhere. Perhaps outside our office is a noisy street with lots of traffic and activity, or we're surrounded by others in a high state of noise, like a high-stress financial trading floor or a room of hungry toddlers. Or we sit down at our desk to find that the Internet connection isn't working, we have 250 unanswered e-mails and four deadlines today.

Our inside frenzy is our internal noise level, partially driven by our response to the thoughts and feelings generated by the outside frenzy.

Whether externally or internally generated, frenzy is a thief. It steals away our sense of being calm, at peace, in charge, in control, the boss of our lives in the moment or overall. So how can we put the thief behind bars and get on with being our best? Here's how:

Awaken to your patterns of calm and frenzy

Dr. Hammerness talked about the importance of regulating one's emotions, processing and managing the negative emotions and harvesting and amplifying the positive emotions. In the case of frenzy, this means being like a firefighter, rescuing your calm out of the fires of frenzy. However, before you regulate, you need to become mindful—aware and awake to your calm and frenzy—noticing moments when you are calm and moments when you are frenzied. When does each state turn up? What triggers them? Are there different kinds or levels of calm or frenzy at work, home or when traveling?

Our experiences with and reactions to calm and frenzy are as distinct and individualized as our fingerprints. One person can feel calm in one chaotic situation, while another will feel anxious and lacking control in that same situation. We need to look to our inner Sherlock Holmes so we can discover which situations allow our negative emotions to take over and better understand our own mysteries along the way.

When calm or frenzy appear on your consciousness radar, what do they look like, feel like, sound like? Find metaphors that describe their essence. When you feel calm, it feels like gentle waves washing rhythmically onto shore. Your eyes relax, you smile and your shoulders drop. You feel grateful and alive. When you feel frenzy, it feels like you are driving in a blizzard, sitting inside a buzzing beehive or surrounded by road rage in a traffic jam.

Start to reflect on your experience of calm. Think about times when you don't notice the frenzy and you feel calm. Recall how you are feeling and what is happening. What can you learn from the calm moments or episodes? Get the thirty-thousand-foot view of both your calm and frenzy patterns by keeping a stress graph.

If you want to get to the thirty-thousand-foot level to view both your calm and frenzy patterns, take a piece of paper and graph out your various life stages, major life events or weekly patterns. Assign a rating to your overall level of calm for each life stage, life event or daily life. Is there a lot of variation—high and low periods driven by life stages or events? Or is there a constant level of frenzy and you can't remember a calm and frenzy-free time?

Let's use a scale of 1–10. Think of 1 as you lying in bed or on a tropical beach: calm, relaxed, tranquil. Think of 10 as being how you feel when the Internet connection goes down in the middle of returning the important e-mail to your boss, and the phone's ringing, and you're supposed to be picking up your kids from soccer practice, and…well, need I go on?

Speaking of kids and soccer practice, keep in mind also—as we take our view from above—that the patterns of stress, which often seem so random and unpredictable in our daily lives, can be more clearly discerned depending on what stage of life you're in. This is part of the value of a stress graph—as you can see in these examples, in which we look at the undulations of our "frenzy" through different life stages and events, and on a weekly and daily basis.

a. Life Stage Stress Patterns

Age/marital-family status/residence/occupation

Twenties	Thirties	Forties to Fifties	Fifties to Sixties
Single	Married/ young kids	Teenage children	Empty nest/ grandkids
Chicago	Milwaukee	Milwaukee	Sarasota
Legal assistant	Part-time legal	Full-time lawyer	Part-time lawyer
Stress rating: 5	Stress rating: 7	Stress rating: 7	Stress rating: 3

b. Life Events Stress Patterns

Getting married	Raising kids	Job promotion	Moving cities
Stress rating: 4	Stress rating: 7	Stress rating: 5	Stress rating: 6

c. Weekly Stress Patterns

Weekend	Workday: Mon–Fri	Work evening: Mon–Fri	Vacation
Stress rating: 3	Stress rating: 7–8	Stress rating: 6	Stress rating: 2

d. Daily Stress Patterns

Mornings (getting kids off to school while trying to get ready for work)	Stress rating: 7
Commute	Stress rating: varies, depending on traffic
9:00am to 10:00am (generally quiet, as meetings and calls don't start until midmorning)	Stress rating: 5
10:00am to 12:30pm	Stress rating: 6
Lunchtime	Stress rating: 5 (subtract 1 if you go for a walk; add 1 if you opt to work at your desk)
Afternoon in office	Stress rating: 5–7, depending on work flow
Dinner (for kids and spouse)	Stress rating: 7 (subtract 1 for takeout, but add 1 back later when feeling tense and guilty for resorting to easy but unhealthy fast food option)
Homework/bath/ kids' bedtime	Stress rating: 7 (subtract 1 when spouse is home to share chores)

Notice in these examples how there are variables within your control that can add or reduce the frenzy in your life. In this example, the working mom whose stress we've graphed could reduce the frenzy in her life by trying to walk more often at lunchtime and by sharing more of the after-dinner parenting chores with her spouse.

Another approach to the thirty-thousand-foot level is to identify the main roles you play in life—boss, parent, sibling, child, friend—and think about the frequency of moments of calm or frenzy when you're in each role. Which role brings the calmest moments, and which stirs up the greatest frenzy?

Role	Rate Average Calm Level: 1–10
Employee	
Parent	
Sibling	
Child	
Friend	

As you become more mindful and aware of your moments of calm and frenzy, you'll be able to "own" your role, your responsibility and your ability to increase calm and decrease frenzy. You may realize that you are truly in a state of burnout—almost always frenzied and rarely calm. Or it may be sporadic—some days, some weeks or some recurring phases of daily life, such as the business demands of month or quarter end.

The more you understand your calm and frenzy patterns, the better you will be at removing or reducing the sources of frenzy, getting in control of your response to external or internal frenzy, or eliciting calm. Define your ideal state of calm when you are in control of frenzy.

When are you at your most calm and peaceful? What circumstances lead to an energetic yet calm state? What's your ideal? If you could wake up tomorrow morning in a calm state, how would that feel? What would your life be like if you felt calm twice as frequently as you do now or if you reduced frenzied moments significantly? Bathe in the dream or memory of calm; let it inspire you to find more of it.

Take care of your mind and body

The quickest path to lowering frenzy is to move your body: go for a walk, go to the gym, take a yoga class or, if you don't have time, do a vigorous physical activity for a short period. Even five minutes of stretching, skipping down a hallway, climbing a few flights of stairs or doing a dozen lunges or squats can reduce the frenzy rapidly.

Feeding your body frenzy-reducing foods also works quickly. Eating enough protein, hydrating with water, savoring a bowl of berries or cutting down on coffee, sugar, processed or deep-fried foods will stabilize your blood sugar and feed your brain with a frenzy-controlling diet.

A few moments of meditation, listening to a favorite song or talking to a favorite friend can also quickly settle a frenzied brain. At the end of the day, celebrating with a walk or glass of wine (just one!) can wash away the day's frenzy. This isn't a book on optimal sleep or stress management per se, but both are critical for reducing frenzy and may be strategies to consider.

Physical sensations of frenzy can be caused by internal inflammation caused by gut allergies to wheat, dairy or other foods. Those who have gluten sensitivity (like me) find that eating something with even a small amount of wheat brings agitation and grumpiness.

Focusing first on developing healthier habits will lower your baseline frenzy, allowing you to more easily address other sources of stress.

Dig up and deal with longstanding sources of internal emotional frenzy

If you start to discover that you've got a level of frenzy that seems out of proportion, illogical or unfounded relative to your external or internal world today, perhaps you have some hidden, hard-to-get-at emotional wounds and patterns that have escaped or overwhelmed your brain's built-in ability to heal. The understanding and reasons for why you feel what you feel may seem out of reach. There are many varieties of therapy available to help you access, understand, appreciate, process, heal and let go of these emotional wounds if you're open to seeking help and figuring out which approach will work best for you.

Perhaps your level of panic or anxiety has biological origins—your brain chemistry is predisposed to overreact or react inappropriately. Talk to your physician to determine whether there is a medication that will help, even in the short term as you work on long-term solutions.

Plan life decisions that will remove big sources of external frenzy

You may have chosen a life path that is increasing your frenzy: the wrong career, the wrong job, the wrong marriage or the wrong social network. These are not easy areas to address, but even deciding that you will address them by proactively taking one small step at a time will provide hope that the life stages will eventually improve or pass.

On the other hand, life stages and events may choose you: problems in a relationship, a bad boss, health issues or the unavoidable process of aging. You may not be able to remove the stressful life situation and so then need to focus on getting stronger and managing the negative emotions a little better, day by day.

Learn that your response to frenzy is a choice, and make a choice to lift the frenzy

The bottom line is that you are in charge, and you have a choice. You can choose to work on being calm more and frenzied less. You can choose to get help. You can choose to bounce back when you have a bad moment, bad day, bad week or even a bad year. You can choose to learn, to grow and to overcome. Calm, not frenzy, is your birthright. It's a treasure that you already have, waiting to be discovered.

Stoke up the motivational fires: make sure you're not *benefiting* from the frenzy

If you're reading this book, it's unlikely that you are in denial about the fact that you have too little calm and too much frenzy. However, even if you are aware and accepting of this situation and keen on reducing your frenzy, it's important to remember that some stress and some frenzy in our lives is normal and may even be beneficial.

It could be argued that without stress there would be no achievements, no productivity, no challenges in our lives. Except for perhaps that week in the Caribbean, getting the daily or weekly stress graph down to 0 is not only unrealistic but unhealthy. Think about all of the good things that you get out of stress—the good job, the wonderful home, and the full and balanced life you've structured for your children. That didn't happen without stress and frenzy. Smile and appreciate how stress serves you. Next time it arrives, embrace stress for its benefits, and let go of excess frenzy, which makes you feel like you're driving with the brakes on.

Experiment

Now you are self-aware of what makes you calm and what detonates frenzy, you are yearning for more moments of calm, you've defined your ideal calm state and you're feeling empowered to get into the driver's seat. You're leaving the driveway, and learning how to drive and how to navigate the road to more calm. Change is not a straight line to the finish; it's a winding road—some steps forward and some, and hopefully fewer, backward steps. So abandon the perfectionist voices and enjoy being a scientist who loves to experiment. Develop some ideas about how you might elicit calm from the frenzy in any given moment. Test them out and see what works best.

Look at a photo that makes you smile. Take a few deep breaths. Think of something that makes you grateful. Send someone an unsolicited note of appreciation for what she brings to your life. Read a comic strip. Go outside and breathe in some fresh air. Do a yoga pose. Water a few plants. Text your wife and tell her that you love her. Take the dog for a short walk. Remember a pleasant memory. Get a colleague a coffee. The possibilities are endless.

Build skills, success and confidence in taming frenzy

Once you've found some strategies for eliciting calm and taming your frenzy that deliver early motivation-enhancing gains, work diligently over a few weeks on becoming skillful and wiring them as new habits. Practice and practice until the strategies become automatic and effortless, until you get to the stage where you can't imagine not actively eliciting calm in a skillful manner when the frenzy bomb goes off. As you get more skillful, you'll get more successful and your confidence will grow like an upward spiral. You'll have more calm moments, which will be very pleasant, even surprising in their impact, and will reignite your motivation to keep going.

You're in control, at least most of the time!

Just as you've become aware of when you're calm and when you're frenzied, stop and notice even the smallest of improvements as you work on having more calm moments and fewer frenzied moments. Score them on a scale between 1 (total frenzy) to 10 (extreme calm), and even if you go from a score of 5 to 6, be sure to pat yourself on the back and feel gratitude for the progress. Share the progress and encourage and help others to set out on this same journey. Calm feels good and is infectious. Spread the word!

CHAPTER 4

Rules of Order / *Sustain Attention*

P AY ATTENTION, GET FOCUSED, be vigilant, stay on task, keep your eye on the ball, listen up, get your head in the game: these are just a few of the many ways we have of calling on that very basic human skill of paying attention.

There may be good reason for the variety of ways in which we ask each other to pay attention. In these fast-paced times, our attentional abilities are highly taxed. The world is demanding our attention at every turn and around every corner.

The ability to sustain focus is one of the building blocks of organization. It is step *two* in our process to help you become more organized. The first step is to establish emotional control—to "tame the frenzy." Now we are ready to take the next step—to sustain attention and to stay focused for greater lengths of time.

Of course, asking people to pay attention is simple. But from a brain science perspective, the actual process of doing so is a remarkably involved task, requiring work from many distinct brain areas. Paying attention is far more complex than, for example, the act of looking at something or someone. Indeed, there are neuroscientists who devote lifetimes to the study of attention. Obviously we can't go into that kind of depth here, but because the ability to pay attention is such a critical component of personal organization, it's important that we understand something about how it works—particularly as our understanding of it has evolved and changed in the past few years.

The first step in what we call the "attentional process" is to orient to the stimulus, whether it's the commercial on television, the teacher at the head of the classroom, or the red light flashing in the distance. Actually, let's imagine that light flashing in the distance is the signal from a fire engine, racing down the street. You turn and look in the direction of that sound, as your brain locks in on it. Think about just that for a moment. Consider how quickly we stop, look, listen; how fast our reaction time is; and how, in the blink of an eye, we have identified what the vehicle is, what direction it's coming from, and its probable purpose. Let's add the whiff of smoke in the air, and you've deduced even further what is happening and where that fire engine is going. In doing all that you have still used three of your sensory modalities—auditory (hearing), visual (seeing) and olfactory (smelling). Your tactile (touching) and gustatory (tasting) senses will have to wait until dinner.

Next step in the attention process is our engagement with that information as the fire engine comes blasting by. First simple orientation to the noise, now lots of attention to detail. You notice it festooned with ladders and tanks, an impressive complement of modern firefighting equipment. You see the firefighters in their gear; you catch a fleeting glimpse of determined faces under their helmets. You read the lettering on the side of the truck and see which firehouse the engine has been dispatched from and recall that you've passed that building. Perhaps you even recall an episode from the television show *Rescue Me,* a scene from the day your child's class visited the local firehouse or something you read in the local paper about the fire department requesting funds for new equipment. You are now attending to this "stimulus" fully, pulling in and synthesizing bits of information from various parts of the brain. You are homing in on the sound and bringing to it the full and awesome powers of sustained, focused attention. And yet it's all happening in a matter of seconds. Take a moment to acknowledge that innate human skill—and to congratulate yourself on your facility for being able to attend to a great deal of information so quickly! Remember also that no matter how disorganized you may feel or how inattentive you may have been on the job or at home lately, chances are that if a fire engine really did come barreling down the street right now, you'd be able to attend to it with the same richness and breadth of cognitive resources we have described.

Let's talk a little about how you are able to do that and about important areas of the brain "cortex," the area we must pay attention to when discussing this. *Cortex* comes from the Latin word for *bark.* It is the coating of the brain surface, those folds/crevices we mentioned earlier—and much of the information processing for the brain happens there. There are several specialized cortical areas, serving different functions.

One way of describing the brain's attentional workings is to begin at the back (posterior) of the brain cortex and move forward, somewhat akin to the way the fire-engine sound began in the background, in the distance, and then became clearer, as the truck raced toward you. Similarly, sensory information taken in by your eyes is received in the back of the brain (occipital cortex) and moves forward in the temporal and parietal cortices.

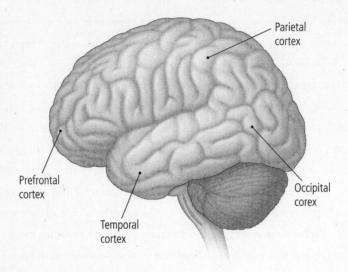

Parietal
cortex

Prefrontal
cortex

Occipital
corex

Temporal
cortex

Source: National Institute of Drug Abuse Teaching Slides

The parietal cortex is involved in scanning the environment around you, analyzing motion, considering spatial relations and orienting you in time and space. It is key in that first orienting step, in which you focus on new stimuli. The temporal cortex is also involved in attending to the features of stimuli. While the parietal cortex gives us the heads-up to direct our attention that something is coming, the temporal cortex allows you to identify more about what that something is—its color, shape, sound and other features. You focus on a particular detail, like the pitch and volume of the siren or other salient information, such as a bright, flashing red light.

All of this information is processed, refined and integrated with memories of relevant prior experiences as the nerve cell messages move forward through your brain, taking information to the "control nodes" of attention in the front of the brain, the prefrontal cortex (PFC)—the area responsible for our actions and responses to this information. The

PFC is considered a "control node" for attentional processes—a critical gatekeeper of attention. This should not be surprising, considering its key role in the management of emotion, as we discussed in the previous chapter. Its function is to help us sustain attention over longer periods of time, as we continue to receive and consider information coming in to us (and plan to start actually doing something about it). The PFC helps us to block out irrelevant stimuli—the cell-phone conversation going on next to you when you first noticed the fire engine or the people walking across the street. While attention begins with orientation and continues with sustained focus, it also depends on our ability to handle distractions that could disrupt us along the way. As we will discuss in the upcoming chapters, the PFC is involved in more complex aspects of attention and organization—including the ability to shift attention from one thing to another and to use memory to keep attention on something, even when it's out of sight.

Of course, in the case of the fire engine, what we have just described is the process and the mechanisms of attention working perfectly, the PFC and other parts of the brain fully engaged and the stimulus powerful and striking.

That's not always the case.

CASE STUDY: THE LIMITS OF ATTENTION

Nancy was in her thirties, a certified financial planner and an up-and-coming star with a local financial firm. But she had a problem.

"I can't seem to get things done. I can't focus on anything. I'm scattered. I'm all over the place," she said, as she sat in my office.

But she had a good job and told me that she's recently been promoted. Surely, she must be doing something right, no? "Yeah, they're happy with me at this firm," she admitted. "But the more responsibility I'm given, the more I'm struggling to keep up. I'm afraid I'm going to blow it."

Because your environment and changes in your environment can affect your ability to sustain attention, I probed her about her job and her workplace. She told me that in addition to her time on the computer or on the phone with customers, her office door was always open, and people were often dropping in with questions. She's also frequently summoned into meetings. Clearly, she's in a busy, potentially distracting situation. Also, it appears as if the level of work, and hence the level of attention and focus required of her, has grown. It could very well be that she's reached her limits.

"Is this something you've noticed for a long time...these problems you're having with focus?"

"Not really," she said, although there have been times, here and there. "In my last job," she said, "I was put in charge of this special project. I suddenly had all this extra responsibility and got so overwhelmed...," she stops and giggles. "One time, I was supposed to show up at a meeting at 5:00 pm, I totally got distracted and went out for drinks and dinner with a friend instead." She looked at me and smiled. "At least it was a good dinner!"

The majority of patients who walk into my office have a clear lifelong pattern of persistent, problematic symptoms of ADHD. The rest are somewhere along the spectrum from "healthy" to ADHD; with symptoms of inattention and disorganization, to varying degrees, for limited periods of times, a diagnosis of ADHD is not made. Nancy doesn't seem to have ongoing, lifelong problems with attention but, rather, sporadic ones at various times in her life. It appears that now is one of those times.

"Do you feel like you've hit your capacity?" I asked her. She furrowed her brow. "My capacity?"

> "Yup," I said. "Sounds like this is a time in your life when "you have reached a certain point, when there is more asked of you, more information for you to focus on, more distractions... like you have reached your attentional limit."
>
> "That's interesting," she said. "I didn't know that your ability to pay attention even had a limit."

A FINITE RESOURCE

Despite all of the brain's impressive attention hardware, there is indeed a limit to what it can deal with and for what duration. How long is a "normal" attention span? The "healthy" adult population can feel confident that their focus can be maintained for upward of an hour—about four times as long as the ten to fifteen minutes people with ADHD often cite as their typical attention span. That is, unless there is an impending deadline, significant pressure from a boss or a spouse or a task that is particularly novel or interesting. In those situations people with ADHD can focus for longer periods of time. But typically, for people with ADHD, instead of focusing on the task at hand, within minutes they are out of their seats, getting a drink of water, looking out the window or surfing the web.

Their basic unit of attention is very brief and has always been very brief, since early childhood.

What affects this length of time? Many factors. Not surprisingly, we tend to pay closer attention to that which is interesting or perceived as relevant and important to our goals or when there are some particularly striking factors related to it. So the meeting with the financial planner about your nest egg; the fast-paced, page-turning novel by the author you love; or the fire engine with the siren screaming, horn blaring and lights flashing are all more likely to be winners in the ongoing battle for your attention. These things are "salient."

Some people's abilities to block out extraneous stimuli and concentrate are legendary: It was said that Ulysses Grant, the famous Civil War general (and later president) had an almost "superhuman" ability to stay focused, even in the din of battle. Cannons roared, smoke filled the air, chaos reigned, and Grant was still able to focus fully on reports from the field and make key decisions. Historian and author Mark Perry called this the general's "most sterling quality. While not the tallest, or strongest, or brightest or even the most insightful of men or generals, Grant brought a singular concentration to everything he did." Grant's remarkable attention powers paid off at the end of his life. Racing against a case of terminal throat cancer, he attempted to finish his memoir in order to obtain for his wife and children the profits of his book, which would resuscitate his family's financial situation. Despite great physical pain and discomfort, he paid full attention to the writing and revisions of the book. He finished on July 19, 1885—and died four days later. Now *that's* focus. A year later, his widow received a check for royalties totaling $200,000.

On the flip side, we have someone like Nancy. Normally, she was probably able to pay attention to her work for thirty or sixty minutes or even more. But when the workload increased suddenly and she was being forced to process and deal with greater amounts of information, the "wheels" of attentional progress screeched to a halt.

Sometimes, if we are not fully allocating and using our attentional capacities (as we might with the fire engine, the good book or the important financial meeting), then the information might as well not be there at all. Perhaps you've had the experience of a colleague saying, "I sent you that memo the other day; don't you remember?" or your spouse insisting that "I showed you where the spare key was; you must have forgotten." You're dumbfounded because you honestly *don't* remember ever seeing the memo or the key. It's as if these things never happened. What *did* happen is that we *didn't* pay attention; we weren't processing that information when we were told or when the memo or key was shown to us. The act of paying attention is not always an automatic process. Without our concerted efforts to do so, events, information and experiences can pass us by. It's intriguing to think that "I can't remember" may really mean "I wasn't paying attention in the first place."

ATTENTION! ATTENTION!

We've been talking about attention as if it's one pure substance or quality. Actually, scientists describe *two* types or modes of attention—goal directed or stimulus driven.

Goal-directed attention is driven from within, voluntarily by our goals and aspirations. This form of attention is consistent with our own unique life, our specific interests or aims of the moment. This form of attention is "top down," meaning that it is rooted in those cortical areas of the brain; the ones that, as we discussed in the previous chapter, are associated with cognitive control. For Nancy, her goal-directed attention can be "on" when she is in the midst of paperwork developed for an individual client, which she sees as critical to foster an emerging customer base in a well-to-do suburb.

Stimulus-driven attention, on the other hand, can be captured by someone yelling fire, a pop-up screen on your computer, a flash of lightning on the horizon or the sound of a power chord on a guitar. Sometimes that information can be life-saving; oftentimes, it is innocuous and arbitrary. This stimulus-driven mode of attention is sensory and external. This may be what is getting in Nancy's way—the external random people and demands that bombard her during the day through her open door.

Scientists continue to debate what makes a stimulus more likely to capture our attention. Perhaps it is the salience—prominence or relevance—of the stimuli or some feature of the stimulus itself, like its sudden appearance? Despite our evolutionary advances, maybe it's still just "shiny metal objects" that grab our attention—as easily as a dangling string attracts the attention of a cat.

Advertisers, who have long studied attention for obvious reasons, understand this well. And some of the most successful ad campaigns are those with messages that are salient to their audience—not just gimmicky commercials that get our attention briefly. One classic example is the early 1980s ad for the Apple Macintosh, with the tagline "The Computer for the Rest of Us"—a line designed to get the attention of many American consumers who were curious about personal computers and what they could do but still felt that these mysterious machines were comprehensible only to those in white lab coats, who held advanced engineering degrees. That campaign helped launch what has become one of the most ubiquitous, successful brands in the world because its message was simple and salient enough to capture attention.

Fortunately, the brain is remarkable in its ability to manage different and competing modes of attention—some of it goal-directed information, which is consistent with our objectives, and some of it stimulus driven, which may run counter to or even change our goals. The optimal balance may be to maintain and develop attentional goals and to allow oneself to be "captured" by only those stimuli that align with our goal at hand.

Consider yourself at a work meeting. While your attention rests on one thing (the speaker at the head of the conference table), your brain continues to evaluate new information (the rustle of papers to your left, the whispered comment to your right) even at a subconscious level. These new stimuli are competing for your attention, but the organized brain is able to evaluate and screen out what is not worthy of your attention. There is still cognitive work to be done: the ability to properly handle all the noise from the environment and evaluate and prioritize it while not being pulled off the main task at hand (listening to the speaker and taking notes) is a basic and important sign of the organized brain. It's one most of us take for granted. For some, however, that's not so easy.

CASE STUDY: SYSTEMATIC DISTRACTION

Jason was a junior in college, and he was struggling. Distracted from his studies, he had watched his GPA plummet. By the time he came to see me, he was in danger of falling below 2.5, jeopardizing his chances to get into an MBA program, which is what he wanted and what his parents were willing to pay for—provided he could get in. "I have one semester to turn it around," he said when he sat down in my office.

"What kind of school are you in?"

He looked quizzical. "What do you mean?"

"Are you at a big university where you have lecture halls with 100 people...or a small school with seminars of five or six gathered around a table? Are you living in a dorm in the center of campus or an apartment off campus?"

All of this was important. We needed to lay out his environment. Turns out, he was in a midsize school in the Boston area and lived in a dorm.

"So where do you study?" I asked.

"The library," he replied. "I'm there for hours sometimes."

"Do you get a lot done?"

"Well...not so much," Jason said as he shifted uneasily in his seat. He told me that, while in the library, he jumps from one assignment to another, taking book after book out of his backpack. He opens them, reviews his notes or reads a page or two and then closes them and puts them away. He watches people walk by. He gets up and rummages through the shelves. He reads the notices on the bulletin boards.

Unable to accomplish much there, he leaves and goes back to the dorm. It happens to be a dorm for upperclassmen, and generally, he told me, it's quiet. "Too quiet," he said. "It almost distracts me. And if there's any noise somewhere...somebody slams a door upstairs or something...it sounds much louder and really throws me off." Eventually, he said, "I just realize I've been doodling for an hour."

In addition, and this is important for the evaluation, I learned that Jason had similar issues when he was younger. Talking about his inability to keep his "nose in a book," even at the library, he chuckled as he recalls his first-grade teacher admonishing him during their reading and writing exercises to keep his eyes on his paper. In middle school, he admitted, his attention wavered as well; in high school, although he did extremely well in some of the classes that he liked, he nearly failed chemistry. "I just wasn't into it," he said. "I'd find myself watching stuff bubble in the lab or reading the charts up on the wall. I think the teacher passed me, as a favor."

Although I am confident that we can help him, by the end of our session my diagnosis is made: Jason has ADHD.

THE SCIENCE OF INATTENTION

The people I see with ADHD, like Jason, struggle with the fundamental skill of attention. The name of his condition, attention-deficit hyperactivity disorder, has changed over time but reflects a fundamental difficulty with the basic unit of attention. In upcoming chapters, we will see that "attention deficits" can be more than just a struggle to pay attention but a struggle to turn attention off as well. Contrary to the popular image, children with ADHD, for example, can spend hours of intense focus on a video game. It's almost as hard for them sometimes to turn that attention off as it is to keep their attention on their schoolwork. Attentional abilities are not just an "on/off" switch. It's about turning attention to the right task at the right time—and then to turn attention off again, when necessary, and in line with our goals.

Studies of individuals with ADHD have found abnormalities in the brain areas that are responsible for our ability to pay attention, including some of the cortical regions we've already discussed such as the PFC. In comparison to persons without ADHD, the brain regions of those affected by the disorder can actually be different in size, in function and in how they are wired or connected to the rest of the brain.

The neuroscience of attentional deficits in ADHD, as well as in other conditions such as stroke or traumatic brain injury, allows scientists to understand more about the working brain. While we marvel at the remarkable ability of the brain to adapt to the modern, ever-demanding world, the study of persons with medical conditions like ADHD remind us that the attentional system can be thrown into imbalance. So while the chances are you don't have ADHD, your inability to focus and seeming lack of attention could very well be real and not a figment of your imagination. As we have seen, we *do* have limits. Fortunately, while that doesn't mean there's something "wrong" with your brain, you are correct in—pardon the pun—paying attention to your inability to do just that. The good news is that there are ways that we can resharpen our attentional abilities, better manage our cognitive load and pull back from the limits we've reached.

ATTENTION! YOUR BRAIN IS UNDER
AN INFORMATION ASSAULT!

The human brain is facing a technologically driven onslaught of information. The advances in brain science over the past ten years have revealed that the brain is a breathtakingly complex organism. It is perfectly capable of paying attention today—and tomorrow. Nowhere is the brain's sophistication more evident than in our attentional abilities—where stimuli and stored information, involving all the senses, are synthesized and interpreted by our brains in seconds. That, of course, is not to dismiss the increasingly complex and information-rich environment that our brains must contend with every day. As noted, attention has its limits; there are times when we may feel overmatched by the speed and volume of all that noise, which is why you may be reading this book. But just as our ancestors' brains had to adapt to new technologies that challenged their ways of thinking—from writing and language to the operation of machines and automobiles—so will we. Trust me, our brains are not going to break down under the strain of one trying to figure out how to work a new handheld device or respond to one more Facebook post.

What research is now telling us is that what "hooks" our attention is usually something consistent with our goals. That's more important than how "loud" or salient the stimulus is. We can process a lot of information about that fire engine, attend to it briefly and then get back on task. But if your cell phone vibrates and you see that it's your spouse, your boss or your physician, well, you're cognitively adept enough to block out the sirens and flashing lights and hook your attention to the phone call, the stimulus that really matters to you. The implication here for someone struggling to stay focused is that we need to foster as much goal-directed attention as we can. We need to be more discriminating and not just go chasing every fire engine—no matter how shiny—that comes racing down our street.

That, we *can* learn to do.

COACH MEG'S TIPS

Many of us have early memories of our schoolteachers chiding us to "pay attention!" We discovered at some point that we were good at paying attention to people and things that were engaging and fun, and we struggled with people and things that were boring or uninteresting. Our parents and caregivers diligently watched for the activities that grabbed and sustained our attention, signs of our individual talents and interests.

As we grew up and got into the full swing of adult life, we found ourselves with such a long list of things to pay attention to that it was easy to lose our sense of our innate abilities to pay attention and our unique preferences of activities and people to pay attention to. Recall the early stages of romance when your sweetie commanded your attention beyond anything else. Remember your best memories and how completely absorbed you were in the moment and place. Your ability to pay attention may be lost but not gone.

The ideal state of attention is called "flow," characterized and studied by Mihaly Csikszentmihalyi (pronounced *Cheek-sent-me-hi*) for thirty years as a professor at the University of Chicago. "Flow," he wrote, "is the experience people have when they are completely immersed in an activity for its own sake, stretching body and mind to the limit in a voluntary effort to accomplish something difficult and worthwhile." The term is used by many people to describe the sense of effortless action they feel in moments that stand out as the best in their lives. Athletes refer to it as "being in the zone." The more flow experiences we have in life, the happier and more fulfilled we are. Paying attention fosters well-being.

In his book, *Flow: The Psychology of Optimal Experience,* Csikszentmihalyi tells the story of a woman with severe schizophrenia in a mental hospital. Her medical team had failed to help her improve. The team decided to follow Csikszentmihalyi's protocol to identify activities where she was motivated, engaged and felt better. A timer went off throughout her day, signaling her to complete a minisurvey on her mood, energy, engagement and so on. Her report showed that her best experience was trimming and polishing her fingernails. So the medical team arranged for her to be trained as a manicurist. She began to offer manicures at the hospital and eventually became well enough to be discharged. She went on to live an independent life as a manicurist. This is the power of paying attention to activities we love to do for their own sake.

It turns out that many of us have most of our opportunities for flow-producing experiences at work, yet we miss out on their pleasure because they are polluted by frenzy (a long to-do list, etc.) or by countless other unnecessary stresses and strains produced inadvertently by most corporate cultures. Sustained attention and flow is a natural state, so rest assured that it is something you can have more of with a little forethought and planning. How can you live a life with more focus, sustained attention and increased flow?

Take an inventory of the times when you're naturally at peak attention

Just as you did with your moments of frenzy, it's important to recognize patterns in your life. Think about your life activities that feel absorbing and effortless and make time fly by and that when you're done, you are energized by a sense of accomplishment. These are times when you are not struggling and when your abilities are stretched slightly by the challenge, enough so that you are fully engaged and interested. Too much of a challenge makes you feel out of control and too little leaves you bored. You are likely applying your strengths liberally, whether you're good at facilitating meetings, playing tennis or the piano, cooking a new recipe or writing a blog. As Dr. Hammerness noted, the research shows us that goal-directed attention is the type that we are most likely to sustain because it is more meaningful to us. Let's make sure we can identify the activities in our life that engage us in that manner.

Do more and harvest more from natural attention-producing activities

Expand your awareness and gratitude for these wonderful moments of flow described in the previous section. Make them even more special by ensuring that they are "clean" experiences, not polluted by frenzy. So close the door, turn off your cell phone and e-mail, and engage and enjoy the activity fully—and with a spirit of adventure, curiosity and discovery. Experiencing these natural attention-producing (or goal-directed) activities—the things that you love to do—in this kind of pure way will show you that there is nothing inherently wrong with your ability to be attentive and that your attention is optimal when the activity is interesting, engaging, uses your strengths and gives you energy and satisfaction.

Initiate moments of mindful attention

Now that you have connected with your many moments of peak attention—and the possibility of many more—start thinking about awakening from chronic frenzy and mindlessness and creating more moments of peak attention. Being present and mindful, enjoying the shower when you're in the shower, appreciating the aroma of freshly brewed coffee and basking in your child's innocent smile are all starting points.

Pay attention to the moment. Take a breath and notice where you are, how you feel, your surroundings, and just be. The skill isn't far from grasp—you had it in spades as a child, with a mind clear of years of clutter, thoughts, emotions and memories. Dig it out and try it on, like a new outfit, every day. Just *be* for a short while. Not do.

The more mindful you become, the easier it will be to initiate your attention and be in charge of your attention rather than allowing it to be out of control.

Explore and apply your strengths: unrealized opportunities for sustained attention

An important lesson from flow research is that we are more interested, engaged and energized when we are using our strengths, while engaging our weaknesses drains our energy, which makes it hard to sustain attention. Other research shows that fewer than one-third of adults have a good understanding of their strengths. So learn about your natural talents and strengths you've cultivated over time. Fortunately, as mentioned earlier, there are many assessments of strengths: VIA Survey of Character at www.viacharacter.org, Clifton StrengthsFinder at www.strengthsfinder.com or Strengths 2020 at www.strengths2020.com. Find a coach or buddy to help you think about your strengths and how you can use them more so that they are at your fingertips when you want to initiate your attention. When I recall that one of my strengths is persistence, it helps me reignite my attention when I am tired or drained. I'm good at persisting when it isn't easy—bring it on.

Convert moments and activities into flow experiences

You can make any moment or activity flow producing or attention sustaining by applying the ingredients that, as Csikszentmihalyi has taught us, can lead to flow.

First identify a goal-producing activity: "Tonight, I will find time to focus completely on my son's state of being . . . not to nag him about things he needs to do but to check in with him and see how he's doing."

Set a goal for the activity: "I will give my son my undivided attention for ten minutes; I will make sure that I listen and resist the temptation to jump in and tell him what I think he should do. I will validate the positive things that he chooses to talk about and emphathize with the tough stuff. I will make him feel respected, loved and listened to."

Look for signs of progress: "As we talk, I'll look to see that my son moves from beyond "yeah," "okay," "whatever," and other typical monosyllabic responses to questions about how his day went, to complete sentences as he realizes that I am listening. I will look for him to smile, laugh, relax and hopefully hug me warmly at the end."

Harvest the result: "I will thank my son for sharing his day's experiences with me; I will offer myself as a resource and sounding board for him anytime in the future. I will enjoy and appreciate the connection we have made during these ten minutes and feel good that it has strengthened our bond as parent and child."

Recharge your brain

Our brain is like our muscles: when we use it too much, it gets tired and needs a rest. After intense periods of focused attention, no more than ninety minutes, take a brain break—take a few deep breaths or get out of your chair and change scenery. If you've been in front of the computer, some gentle stretching or a short walk will do wonders for your body, not to mention your brain.

Pay attention to small activities

Much of the above discussion has focused on longer activities because that's where we get traction and make forward progress in our lives, building our families, friendships, careers and well-being. That's what builds our satisfaction with life over time, like a foundation of well-being to stand upon that grows with life experience.

Paying attention matters in the smaller moments in everyday life, too: driving to work, short conversations by phone, clearing a few e-mails, cooking a quick meal, folding laundry. Pay attention and appreciate how full small moments can be. Find the special in the ordinary. That "stop and smell the roses" advice we've always heard may sound trite, but there's truth there.

It's harder to be mindful when we're swimming in a soup of negative frenzy. Yet negative moments are part of being human and deserve our full attention. Don't distract yourself totally from the negative. It is often said that we learn more from our failures, disappointments or setbacks, provided that we are receptive to the lesson.

Here are some examples of how you can find the silver lining:

Your flight is delayed, and now you're stuck for two extra hours in the airport.
Use the time productively: read a little more of your book or use this opportunity to crack open a new book you've wanted to start but didn't have time for. Or perhaps you didn't have time this morning to exercise—well, you can use this time to walk around the terminal or do some yoga in the meditation space in the airport lounge.

You have a fight with a colleague over responsibility on a project.
How can you use this to build a better relationship with this person? Perhaps this is an opportunity to restructure some of the ways things are done at your workplace. Maybe you can work together to present to the boss a new plan of doing things?

Your spouse is ill, and for the next few weeks, you will have to take over his or her responsibilities.
This is a way to better appreciate your spouse's contributions in your life but also to challenge yourself. Now you'll be paying the bills...

or doing the laundry . . . or fixing the meals for a while. These are new and valuable domestic skills to acquire.

You missed an important project deadline.
It could be something like this that prompted you to pick up a book like this in the first place! This is a wake-up call to you—you need to get better organized and on top of things, which you are now well into the process of doing.

Learn how to handle distractions

While we can minimize distractions by closing the door, working in a quiet room and turning off the cell phone and computers, distractions will arrive despite our best efforts. First, listen to your needs: perhaps you are presently easily distracted because you're tired or overworked or you are not in a mindful place or the task before you has been badly designed to be boring or anxiety producing. Your susceptibility to distraction is a sign to pay attention to. What is it telling you?

Mindful attention works beautifully when a distraction arrives, knocking at your door demanding your attention. It allows you to make a conscious choice in the moment without creating frenzy—choosing to switch attention to something new or choosing to notice it and then tune it out. Getting control of one's choice of response is the real lesson here. You're in charge of your attention; you make the choice to follow the temptation of distraction. Get into the driver's seat.

The story about the television star's illicit romance pops up on the computer screen—do you really have to read that now?

You receive a call from your friend wanting to chat or the drop-in visit from a colleague who wants to gossip—can you politely tell them when they ask the perfunctory "Got a minute?" question that no you don't, but you would like to talk with them, and you'll get back in touch in a little while?

It's really not that hard to say "no" to distraction.

Train your mind

It may be that your brain has become wired into a state of frenzy and chronic distraction and, like a muscle group, needs some training to find the old wiring for your capacity to focus. Learning how to meditate is all about learning to pay attention to the present moment and may be one of the best investments you can make. Or coach yourself by setting goals for paying attention, and pair up with a buddy for accountability. How focused were you today? Rate yourself between 1 and 10. Focus on increasing your score slowly and surely over time.

Let your mind wander!

As Dr. Hammerness described in the first part of this chapter, attention is an innate, human ability to be tapped into and optimized. It's not like learning to walk but rather a skill that can be sharpened. Being attentive is critical to your becoming more organized, but let's not forget that there are times when it's good to be distracted, to let your mind wander. You can and should practice and work to become more attentive in many situations of your life—but not all. Make sure you do take time to turn off the switch once in a while. Enjoy the random thoughts and shiny objects.

CHAPTER 5

Rules of Order/*Apply the Brakes*

DEBORAH WAS FRIENDLY, CHEERFUL AND SELF-ASSURED. In her mid-thirties and the happily married mother of two children, she and her husband had a house in the suburbs of Boston, with a big backyard and a great home entertainment system. She was the class mom at her children's elementary school, she drove the kids to their soccer matches and music lessons and she tried to make healthy meals and keep the house warm and welcoming.

And to round out this picture of suburban bliss, Deborah and her family had a cute little terrier named Snickers.

What, I wondered, brought her into my office?

"I'm having trouble getting things done," she explained, apologetically.

No need to apologize, I told her. And it seemed as if she was doing a pretty good job of raising a family based on the smiling wallet photos I'd just been shown. Even Snickers looked content. What exactly was going on?

Deborah calmly explained the problem she was having.

"I'm struggling to get the things done that I need to," she said. "I start out planning to do A-B-C, and I never get past A."

I asked her for an example.

"I've got a doozy," she said, rolling her eyes. "Last weekend, we were going to clean out the garage."

Alarm bells rang and red lights flashed in my mind. The garage! Uh-oh. I'm not sure why this is—although I am certain someone could probably earn a doctoral degree finding out—but I hear about the garage a lot in my line of work. Whenever I do, I know there's going to be a problem. People seem to go out to the garage and never come back—or at least not in the same frame of mind.

I listened, wondering what the insidious garage had done this time.

"We went out there to organize it," Deborah said. "And we had tons of old toys and sporting equipment and some of my husband's tools and boxes. Let me tell you, there was a lot of junk out there."

Sounds like most garages. Probably a good idea to get it sorted out. What happened?

"Well, let me just say I'm a person who likes to finish what I start," Deborah said and then looked at me quizzically. "That's a good thing, isn't it?"

I nodded, hesitantly. "It can be...."

She went on. "So we started after lunch, I think it was about one o'clock. I said I'd work on it for an hour, and my husband was going to help."

At 3:00 pm, Deborah was in the garage.

At 4:00 pm, Deborah was *still* in the garage.

At 5:00 pm, yup...you guessed it. In the garage.

Deborah ended up being out there for over four hours. She said she had gotten caught up in the cleaning and organizing process—and just didn't stop. She looked through some of the stuff she was supposed to be throwing out: she began reading old letters, inspecting old clothes, perusing old books. Then she decided to pull apart some of the shelving that she'd noticed after rummaging through some boxes. She saw that the garage needed sweeping out. She climbed up to investigate what she thought was a squirrel's nest in the rafters.

"I'm a go-go type," she said, as she recited the list of what she'd accomplished in the garage. "Once I get going, I can't stop, especially when I see something else that needs to be done."

That's fine, but not when it means that a lot of other important things don't get done. Other appointments, other chores, other parental duties...all, she admitted, had been forgotten during her extended sojourn in the garage.

She even forgot to feed Snickers.

Had this happened once, or even once in awhile, it wouldn't have been an issue. But, Deborah acknowledged, this kind of thing was happening "all the time."

And what exactly was the problem? Let's walk through our Rules of Order up to this point.

Rule 1: Tame the frenzy.

Okay, here, Deborah seemed to be doing well. She was not emotionally distraught. Certainly she was annoyed and a bit frustrated by the fact that she couldn't seem to get from A to B, but in terms of her overall emotional state, she seemed quite stable and in control.

Rule 2: Sustain focus.

Again, Deborah seemed to be doing well, meaning she could stay focused on one activity: organizing the garage.

Her problem involved our next "step."

Rule 3: Apply the brakes.

Applying the brakes means exercising "inhibitory control." That has nothing to do with being inhibited, which in the common use of the word refers to a sense of repression, an inability to express feelings openly or behave spontaneously (as in, when a shy child is *inhibited* in the presence of a strict, disciplinarian teacher).

Inhibition, as we refer to it here, means an ability to restrain or regulate or control your attention. Those individuals who have difficulty doing this, who find it hard to stop some activity that is no longer effective or productive, just plow ahead without stopping to think (such as people with ADHD). It was telling that Deborah labeled herself a "go-go" person. One of the common cognitive tasks used by psychologists to study inhibitory control is called the "go/no-go" task, in which a subject is asked to respond to "go" signals and not to respond to "no-go" signals.

Deborah needed to exercise a little more "no-go" instead of just "go-go"—as do we all if we wish to become better organized in our lives, both at work and at home.

NOT INHIBITED BUT IN CONTROL

Your ability to apply the cognitive (or physical) brakes—to thoughtfully "inhibit" an action that may lead you down a rabbit hole of trouble and confusion—is a hallmark of an organized mind.

It's akin to the importance of a good set of brakes on a very expensive car. In Deborah's case, it would have been the ability to stop, think and walk away from a not-yet-totally-clean garage instead of attending to more pressing tasks. Inhibition allows us to be adaptable and to stop behaviors that are not needed and, in so doing, further supports our ability to stay organized and on top of our game in the face of a changing, evolving environment.

Researchers who study ADHD and cognitive problems consider inhibition as a self-regulatory act, critical in successful functioning. ADHD guru Dr. Russell Barkley has written that the inability to regulate oneself is an essential characteristic of the disorder. He elaborates on the ways that this can be seen in those with ADHD:

> Impaired response inhibition, impulse control, or the capacity to delay gratification . . . is often noted in the individual's inability to stop and think before acting; to wait one's turn while playing games, conversing with others, or having to wait in line; to interrupt their responding quickly when it becomes evident that their actions are no longer effective; to resist distractions while concentrating or working; to work for larger, longer-term rewards rather than opting for smaller, more immediate ones; and inhibiting the dominant or immediate reaction to an event, as the situation may demand.

Even without ADHD, the inability to apply the cognitive or physical brakes can be seen in various aspects of our lives as well. Although Dr. Barkley is focusing on the behavior of children in that passage, adult examples of this lack of inhibitory control can be strikingly similar. Think about the guy at Dunkin Donuts who tries to cut to the front of the line in order to be served immediately, those who constantly interrupt conversations because what they have to say has to be addressed now, or the drivers who like to lean on the horn in traffic or who, rather than sit patiently at a light, will impulsively wheel their cars around and find an alternate route—one that may ultimately cost them more time than had they simply waited for the light to turn green—simply because they couldn't control the urge to keep moving.

Let's face it (and most of us have had to learn this from hard experience): sometimes the most important action is nonaction. Stand in line for a minute or two, and you'll get your hot cup of coffee and muffin anyway. Let other people finish their sentences, and they might be more likely to listen to yours. Wait for the light to turn green, and you'll probably get home faster. Our ability to resist the competing demands of the world around us, to regulate our responses to them and to delay gratification—this is another key building block of success and of organization. Yet many of us keeping falling into the same trap. As Deborah's long afternoon in the garage demonstrates, even with a cool head and sustained focus we can be left with the most important tasks undone if we don't learn how to stay on task. It can feel puzzling, as it did to Deborah, given her clear ability to accomplish things. How can it be, that a person like her—or maybe like you—can be disorganized?

To find out, let's deconstruct inhibition—or "nonaction"—from a scientific perspective. As noted above, cognitive research suggests that several processes may be at work in effective inhibition. Keep in mind that our Rules of Order are like building blocks, laid one on top of another. One of the most important of these processes should be familiar from our last chapter: the ability to pay attention and handle distractions—that ability to block interference from irrelevant stimuli (remember the stimulus-driven attentional system we talked about in the last chapter?) in order to keep our goal-directed attention functioning well. Our ability to apply the brakes of our car begins with having brakes that are lubricated, balanced and ready to go. An effective pair of cognitive brakes keeps you prepared for distractions and ready to take them in stride. In so doing, your interference control system is primed and ready to be able to do two things:

- to inhibit what is considered an obvious or expected response to a stimulus
- to stop an ongoing response

These are two key aspects of good inhibitory control. Let's look at each a little more closely.

In the first situation, you are challenged by not reacting in a situation in which you might have reacted in the past, a situation in which it might seem quite reasonable to react. Maybe a friend calls on the phone. "Hi," she says. "I'm stuck here at work, and I'm wondering if I could ask you a favor? Could you pick up my daughter from her soccer practice?" It seems reasonable to react with an automatic "yes" and grab your keys. But the organized mind pauses and thinks for a moment and then inhibits the instinctual "I'm on my way!" action and reminds your friend that today is an extended practice and she still has an hour to get there, thus saving you both a lot of time and stress. This is exactly the sort of "small" issue that can happen repeatedly throughout the day and week when inhibitory control is not working well. It can lead to wasted time, frustration and unnecessary stress. It can take a toll, like it has for Deborah.

In the second situation, you are again challenged, but this time you're in the midst of doing something. You are acting or moving, and you must come to an abrupt halt. Let's say you're at work, filing some important papers, when a new colleague comes in and ask one of those irritating but strikingly insightful questions, which often only come from a new set of objective eyes in the office. "Wow, you're still using a filing cabinet?" the newly minted college grad asks perkily. "That's, like, so 1999. I'm surprised you don't have a totally paperless office."

The organized mind is quickly on top of this situation. Brakes are applied to a possible response questioning the age and maturity level of this colleague—or to snarky remarks to the effect of "Why don't you go listen to Lady Gaga and let the adults do the work?" Inhibited, as well, are defensive comments such as "This is way I've been doing it for years, and it's worked fine. So what's your problem?" No, the well-primed cognitive brakes bring that to a screeching stop. Instead, the organized mind remains calm, rational and focused and stops your "so-1999" activity. You consider that the person has a very good point; this activity really is inefficient and is perhaps no longer the right approach for the office. Maybe it would indeed be a good idea to learn how to scan documents and make an ally in the process, so you suggest something mature and savvy like "You know, with my experience and your new eyes on the scene and tech know-how, we can make some real improvements here. Let's meet next week about how we can make some changes."

Does some of this just sound like exercising good judgment—the admonitions to "bite your tongue," "keep your eye on the ball" or the equally old adage to "look before you leap?" To some extent, yes. Those hoary words of wisdom instinctually recognized this complex process that modern neuroscience labels "inhibitory control" and also recognized the fact that there are lots of times in our lives when that kind of control must be exercised. But however you verbalize it, it's a real human phenomenon that is very relevant to how we live and function and organize ourselves on a daily basis.

How important? In a recent study by a well-known group of researchers in the Netherlands, including the eminent Joseph Sergeant, a series of cognitive tests were presented to adults with and without ADHD. The greatest difference between groups was seen in tasks reflecting inhibition; adults with ADHD performed worse. IQ didn't matter. Gender didn't matter. This study supports the idea of ADHD as a disorder of self-regulation. Considering again that our contention in this book is that we have much to learn about organization from people with ADHD—those who can struggle mightily with disorganization—the idea that inhibition may be *the* fundamental deficit in ADHD makes this an important process to consider in our pursuit to be organized.

However, the roots of inhibitory control are less understood. A recent study in *The American Journal of Psychiatry* by a Canadian research group found that poor inhibitory control in children can be predicted by the abilities of their parents. At this point, we don't know the influence of genes or the environment on deficits in inhibitory control—is it genetic or can you "learn" poor organization from the environment and from people around you? It's an interesting question, and if those who seem to lack the well-lubricated cognitive brakes choose to blame it on their parents, go ahead. The truth is, regardless of the cause of your weakness in inhibitory control, it can be improved, as suggested by yet another recent study involving people with and without ADHD.

In this study, subjects were asked to perform a series of tasks that tested inhibitory control. The researchers then used neuroimaging to look at the brain activation patterns of the adults with ADHD, as compared to control subjects. While performing tasks, the subjects with ADHD used alternate brain activities to compensate for their inability to regulate. The implication here is that even for those with a disorder such as ADHD, the brain can perform, or at least muster the effort to attempt to perform, tasks involving inhibitory control.

THERE'S NO STOPPING THESE STUDIES

No one seems to be putting the brakes on the study of inhibition. It has really emerged as a focal point of research as more scientists and mental health professionals begin to realize its importance in understanding a whole host of healthy human behaviors—including our topic: the ability to stay organized.

So how do the scientists study this process? They use tasks that involve applying the cognitive and behavioral brakes. Two common tasks have quite appropriate, intuitive names: the "go/no-go" and "stop-signal" tasks.

In these tasks, subjects sit in front of a computer screen and have to respond to noises or pictures that emit and flash in front of them.

Simple instructions are given to isolate a specific brain function—such as sustained attention or inhibitory control.

In the stop-signal task, subjects are asked to identify a target as quickly as they can but then to cancel or inhibit their response with a so-called stop signal. Stop signals happen randomly, and as you can imagine, the closer one is to the point of no return—your hand is on the button, you are milliseconds away from pushing down on it—the harder it is to inhibit that response. Healthy persons usually need about 200 milliseconds of lead time in order to stop the response. So to subjects taking the test, the commands come across as "respond…respond…respond…respond…STOP!" Perhaps it's not unlike the boss who says "yes, that's right, that's good, excellent, thanks," and then suddenly, "no, that's not it at all, you've done it all wrong!"

A similar test of inhibitory control is the go/no-go test, in which subjects respond quickly to specific "go" target letters on a computer screen and don't respond or inhibit a response to alternate "no-go" letters. These tasks assess one's ability to stop, to inhibit a response—whether it's in the process or has not yet begun. Think of the game "Red Light, Green Light" you used to play as a child. You would line up with a bunch of friends, and then someone would shout "Green light!" You would all walk forward as fast as you could until you heard the command "Red light!" and you would all try to stop on a dime. Perhaps you can remember some kids falling over, twitching and giggling as they tried to "freeze" and remain perfectly still. Whereas others reacted instantly and stood still as statues—and still others disregarded the signals altogether and forged ahead.

These cognitive tasks are similar to this classic childhood game—and the true victory is won by the person who can *stop* as quickly and effectively and efficiently as they can *go*.

Scientists use these cognitive tasks in concert with the latest in neuroimaging tools to test a person's ability to stop, thereby learning about the brain activity that is responsible for this skill. Studies using a host of neuroimaging techniques have identified specific brain networks as critical in inhibition. Signals can move from one brain region to another—such as the frontal cortical region and the basal ganglia—in order to coordinate a successful "stop" response. Research is also considering how other factors, such as motivation or emotions, might influence inhibition.

Some researchers have likened this process of inhibition to a horse race. Imagine two horses jumping out of the gate and heading down the track. One is carrying the "go" signal, the other the "stop" signal. The "race" (the neural processes) is triggered by a possible "stop/go" situation. It's 6:00 pm; you're about to head home, and the phone in your office rings. You don't recognize the number on the caller ID. Should you pick it up or not? It's tempting. This could be something good…but on the other hand, you promised your spouse that you would stop on the way home to pick up some things for dinner tonight. Could it wait until tomorrow? Or must you pick up that phone *now*? The race inside your brain is on! The "go" horse takes the early lead as your hand begins to move to the phone. But another nerve circuit—the "stop" horse—is gaining. Each bears a different message:

"Go…because you really want to know who's calling, don't you?"

Or:

"Stop, because if you get embroiled in a call now, you'll mess up your family's dinner plans, and you can still get the message tomorrow."

They're neck and neck; who's got the most cognitive juice to win this race? They come around the backstretch; it's going to be close. And the winner is…

Well, the winner is up to you. But in a sense it is a competition between brain signals racing through a circuit, between inhibitory control and no control.

If you want to get better organized, you must learn to obey the "stop" sign. But life is complex. Stopping at one time may be easier than another, depending on the context. For example, it may be easier to say "no" at work but not socially—or it may be influenced by your emotions at the moment, in which case frenzy can dictate the winner. So don't beat yourself up if you make a "go" decision that you realize later should have been "no-go." Success in inhibitory control is not a one-horse race; it's how you manage it in the long run that counts.

CONTROLLED SWING

Let's take these cognitive tasks out of the controlled environment of the lab with subjects reacting to prompts while having their brain activity imaged. Sure, you push buttons at work all day and people push your buttons at home, but life is more complex than a series of "stop/go" signals, isn't it? You need to be able to inhibit more than a prompt on a computer screen, yes?

One group of researchers at Arizona State University examined inhibitory control as it applies to a more complex task: swinging a baseball bat. Think for a moment about the brain skill needed when you're up at the plate and a pitcher winds up to fire the ball in your direction. The fastball seems to be coming right over the plate. It looks like a pitch you could knock over the fence. Do you swing? Not so fast. It could be that the ball is about to tail off low and out of the strike zone. It could also be that this pitcher may throw high. Here comes the ball: do you swing or "stop swing"? Or perhaps you "check" swing—meaning a halfway swing that you try to stop in midmotion once you determine that the pitch is out of the strike zone?

This interesting study, which examined what the researcher called this "complex, multistage" act of swinging a bat—and the circumstances that could lead one to stop fully, partially or not at all—reminds us that everyday life is a bit like being up at bat. It's rarely just "stop" or "go." It's more like a long at-bat. There are check swings; there are hits and misses. You hesitate; you evaluate the situation and quickly assess the pros and cons of action versus inaction. It's also an evolving process; sometimes, new information becomes available. For the batter, it could be the motion or a look from the pitcher, a breeze that suddenly picks up or the signals flashed by the third-base coach. We're all getting such signals that have to be evaluated as we decide "go" or "no-go."

Here's an example: the phone rings in your office, and as you reach to answer, a colleague passing by says "Don't pick it up. It's that pain-in-the-neck client from Acme Diagnostics. I know why he's call-ing...he just rang me, as well...and it's nothing you can help with." You freeze. Should you pay attention to this signal? Why is that col-league telling you this? Is he certain it's the same caller? Do you pick up or not?

The batting experiment reminds us that inhibitory control is not necessarily an open-and-shut case—or to use baseball terminology, a case of balls and strikes. We need to pay attention to the signals. They may be complicated, but you can learn from them. We also need to be adaptable as circumstances change. Just because we chose "go" the last time in this situation doesn't mean we should the next time around.

Stop. Go. Answer the phone. Let it ring. Swing at the pitch. Don't swing. Check swing. Apply the brakes. Hit the gas. However we try to conceptualize it, the underlying concept—that your ability to know when to avoid jumping off task, to hold back, to temper the action with thought and to avoid a ready-fire-aim response—is critical to getting through the busy days of your life.

Let's find out now how to do it better.

COACH MEG'S TIPS

When we succumb to our impulses, we are allowing our emotions to respond unchecked to a request for our attention and then to drive our behavior without stopping to think about our options and make a thoughtful choice. While our brain machinery is complicated, a simple formulation is that we are each unique when it comes to our natural and learned abilities to catch our impulses, activate a thinking process, appreciate the signals our emotions are sending us and make a conscious choice of what to do about this request for our attention.

The field of emotional intelligence has taught us much about how to self-regulate or self-manage our emotions—how to get into control, learn and direct our emotions in the most productive fashion. People vary widely in their abilities to be aware of and manage their emotions. Those with a lower level of ability will succumb more often to impulses, making an unhealthy choice like eating too many cookies, getting sidetracked by phone calls or texts, or lashing out prematurely or angrily at a slight by another.

Our ability to put on the mental brakes and regulate our emotions and behaviors also declines when our energy flags or when we are tired, hungry or stressed out. It can evaporate completely when we are gripped by strong emotions or feel helpless or vulnerable. So what can we do to put our brains in charge, honor our emotions and make the best choices?

Allow thinking and feeling to work to together

Earlier in the chapter, Dr. Hammerness used a horse-racing metaphor to describe the "go/stop" circuits of the brain. We'd like to return to the track to underscore an important point about how you can help manage those circuits and impulses to become better organized.

At our best, our thoughts and emotions work together like a well-trained team—like a world-class jockey and his horse. The jockey (thinking) is exquisitely present and sensitive to the needs of the horse (emotions) and how to make the most of the animal's talents. The horse feels the respect and sensitivity, doesn't take charge or rebel, and responds beautifully to support the jockey's drive to win a race.

When we overthink—when we ride or apply a lead foot to the cognitive brakes—we are like the jockey who holds back on the reins, denying his horse's desire to run and run fast. When, on the other hand, we allow our emotions to control us, without cognitive brakes, we are letting the thoroughbred run wild without the guidance and control of the jockey. Our thoughts, like the jockey, must stay in the saddle and work with the emotions in a sensitive, kind, respectful way, while sometimes asserting a firm hand, reining in the horse and applying the cognitive brakes.

Imagine a conversation with your teenage daughter about the colossal mess in her bedroom. She's upset because she has lost her cell phone, which is not surprising given the clutter and disorder in her room. You've lost count of the gazillion times you have asked her to clean it up. You walk into her room today and face not only the mess once again but also your daughter crying about the lost cell phone, which she swears she had last night when she came home. Your emotions heat up and feel unmanageable. You are about to unleash a stream of vitriol. "How many times have I told you to clean up this room? What are you thinking? Of course you can't find your phone in this chaos! How can you ever find anything? Meanwhile, you have no problem bringing in new stuff, new clothes, to add to the mess. I can't stand this irresponsible behavior anymore."

But hold on—pull back on those reins. The jockey—your thinking brain—now enters the process and steps on the brakes. Here, your brain needs to have a quick chat with your emotions. "Yes, we're very frustrated, and yes, we did tell her so. But all this yelling hasn't worked in the past and isn't likely to work now. It's only damaging the relationship with our daughter. She'll go off to college soon...do we want her to remember us as loving, supporting parents or as constant nags? So let's hold our tongue and instead help her locate the cell phone. And maybe at some point later, when everyone's cooled down, we can revisit this issue with her."

Be careful about applying the brakes to impulses without acknowledging the emotions and sending them elsewhere in your brain to sit and fester. Don't try to be a robot. Those impulses aren't your enemy; they are your teammate, like having a colleague who is creative and spontaneous while you are prudent and self-controlled. Like that colleague—or like the jockey and his horse—you need each other for optimal success.

Don't be afraid to have a heart-to-head conversation

What the jockey really can't do is have a conversation with his horse. But your thinking brain and your emotional brain can and should have a brief tête-à-tête, especially in a potentially volatile situation.

Let's go to Dr. Hammerness's favorite place—the garage. You need to clear it out, and you've allocated sixty minutes because you have other things to attend to. But once you get involved in the job, you realize just how long you've neglected this chore and how much stuff there is out here, and your heart sinks. And as it does, it sends a message:

Heart: I feel sick and tired about the mess in the garage. I really want this feeling to go away, so I'd like to just keep working out here and clean it all up so I'll feel better.

Head: I'm sorry to hear that the mess is distressing, and I empathize with your desire to get it over with once and for all. However, we must be realistic: it will probably take several more sessions and many more hours to complete the job…and do it right. So it seems to me that it would be best to stop after one hour, feel good that we've made some real progress out here, and then move on to the other important tasks we've planned for today. That way, by tonight, we'll feel good about our overall accomplishments and not just be miserable because the other things were neglected and the garage *still* isn't where you want it to be.

Heart: Thank you for your wise counsel, oh brainy one, but I don't think I'm going to feel good about anything if so little is accomplished in tidying up the garage.

Head: Now, now. Let's not get snippy. How about this: I'll remind you before bedtime that we had this conversation and that we made the best decision? By then, we will have made a small but meaningful dent in the garage…and the other tasks will have gotten done as well. In a way, we will have accomplished even more. Think about how good you'll feel then!

Heart: You make a good point. Okay. Let's tell the lungs now to take a deep breath, and I'll leave the garage and move on to the other stuff we have to do today. And hey…I'm glad we had this little chat. We should do it more often.

Know yourself—and apply that knowledge

As you can probably tell by now, one of the best ways to help get better organized is to become more aware of the situations and circumstances in which we are not. Seeing patterns, whether they're with our sense of frenzy, in our ability to sustain focus or (in the current Rule of Order) to apply those brakes, is a key step toward starting new habits.

Keep a journal for a day or two and make a note when you respond to an impulse without thinking, without braking. What was the trigger? What did you do without thinking? Did you think a little but not enough? The time spent noticing how impulsive you are and when and what led you to ignore the impulses is a first step to training your brain to better manage your emotional impulses.

You may notice that you do a great job managing your impulses with work colleagues but fall apart during a tricky conversation with your spouse or child. Or you may see that some colleagues seem to be able to activate your emotional impulses instantly and inactivate your thinking brakes by implying that you are underperforming in some way. Or as we say, they know how to "push your buttons." Instead of pushing back, maybe you learn to apply the brakes instead. Other patterns may emerge: you may be able to exert your inhibitory control better on Monday morning than Friday afternoon. Being aware of that may help you next Friday!

Remember, your ability to manage your impulses isn't a constant, and being curious about the variability will help you find ideas to improve and situations to work on.

Manage your energy

Appreciate that it is harder to apply the cognitive brakes when the car is out of gas, when you're driving on empty in terms of physical and mental energy. Typically we are best at regulating our thoughts and emotions early in the day after a good rest and breakfast. Our worst periods are at the end of the workday when our brains and bodies are depleted.

Your brain doesn't have a way to store energy, which means that if your blood sugar is depleted so is your brain. Keep your blood sugar at a steady level by eating lean protein with every meal and snack. Avoid high doses of carbs, which lead to a spike and crash in energy.

Personally, I haven't discovered a better way to process and tame my impulses than to exercise; whether vigorous or not, even five minutes can calm down my emotions and make it easier to handle emotional impulses. Next best is a good night's sleep, allowing the brain's machinery to do its work in processing emotions. There are studies that confirm the energy-enhancing, emotional-calming effects of both physical activity and sleep.

When in doubt, take a walk. Or sleep on it and tackle it again in the morning, when you're well rested.

Apply the lessons of crisis in daily life

Remember the famous pilot who landed a crippled plane in the Hudson River and saved 155 lives? Imagine what it was like for US Airways captain Chesley B. "Sully" Sullenberger on that frigid morning of January 15, 2009. Think about how many impulses he had to manage and how he had to put the brakes on paralyzing fear and the terrifying thought that his life and the lives of his passengers and crew were in grave danger—and then summon himself to think clearly.

Talk about being able to tune out distraction! Sully showed us all that it can be done in the most extreme circumstances.

In fact, sometimes the extreme impulses call us to be our best selves and we are at our peak in terms of handling our impulses. While Sully may be one of the most famous recent examples, many of those who handle emergencies—doctors and nurses who work in emergency-room wards, firefighters and police officers, soldiers and those in command—are outstanding models of impulse control.

But while you might not have the coolness to land a plane safely in a river, command a platoon in combat or work in an emergency room, don't sell yourself short. If you've been through a crisis or a period of danger, you probably noticed that during those moments or hours you became a little more adept at handling impulses. Many of us do—we just haven't been tested. Not that we're looking to put you or anyone into such a situation. The point is that you may be better at impulse control than you realize. It could be the daily "minicrises" of day-to-day life: a crying child, a rude store clerk, an inconsiderate colleague. These are all crucibles for developing impulse control. Rise to the occasion. I know you can do it.

Develop your own way of braking

Years ago I coached a woman, a freelance journalist, who was struggling with anxious outbursts caused by her relationship with her overbearing boss. She would sit down to write an article with a tight deadline. She'd make a good start and then she would get a call from her boss, who was typically in an agitated and impatient state. For an hour or more after his call, her emotions would interrupt her task as she thought about what she wished she had said to him. Her agitation would steal her attention so that she wasn't productive or creative, making her feel like she was in a sailboat tossed around by waves of frustration, not getting anywhere.

Although she didn't work for a broadcast network, we created what we called an "ABC" process where she worked on practicing three steps.

1. *Awareness:* Become a "kind witness"—and not a stern judge—
    to the arrival of the impulse to call him back and give him
    a piece of her mind. (In other words, I didn't want her to
    beat herself up over the impulse to give this guy a piece of
    her mind!)

2. *Breathing:* Take a few deep heart breaths (moving her awareness to her chest and heart) to bring in empathy and acceptance of her frustration.

3. *Choosing:* Make a conscious choice to apply the brakes on the impulse to call him back and the nonproductive rumination that it created, while promising to give her frustration more attention later. She eventually made time to write her boss a letter that explained her needs and frustrations, and she requested a new relationship dynamic. That launched a series of conversations and a more productive relationship.

Other tools used in the training of emotional intelligence are quite similar: the STOP tool—**S**tep back, **T**hink, **O**rganize your thoughts, then **P**roceed, or PRO—**P**ause, **R**elax and **O**pen.

No matter what acronym you apply, the process is the same. Instead of grimacing or cringing when the impulsive interruption arrives and your emotions take over, welcome your emotions. Try to understand what they are trying to say to you. Thank them for registering an opinion and for taking a stand. Then decide what's best for the team.

Practice the fine art of emotional balance

I like to think of people as leaning toward one of two camps much of the time.

The first one is Camp Spontaneity: Here we live in the moment. We're spontaneous and creative; we allow our impulses to drive us. Want to go skinny dipping in the lake? Sure, why not? Roast marshmallows by the fire and stay up all night? Let's do it! We indulge in the invitations around us, unconcerned about the future.

On the other side of the lake is Camp Sobriety: Here we look to the future. We conserve our firewood; we look before we leap. We are like the industrious ants as compared to our live-for-today grasshopper friends at the other camp. We are happy to forgo the instant pleasure of following an impulse for the opportunity to feel satisfied with our accomplishments down the road.

Most of us spend our lives between both camps. We tamp down our impulses much of the time and then once in a while we have the strong urge to be impulsive about something. The key is to welcome the impulse and decide whether and when to indulge it. Have an ice cream cone once in a while, steal away from work for a couple of hours to meet a friend or spend a little extra to get a special sweater.

Let go of the brakes—unless of course (and this is the thinking part of the brain talking) your doctor has put you on a strict diet, your impulsive get-together with your friend means you have to miss a very important work meeting or your financial pinch is significant enough that you really don't have that little extra to spend on the sweater.

The impulse to visit Camp Spontaneity often comes knocking at the door of your cabin. Come on in; the water's fine! Don't worry, be happy. Enjoy a treat, do something special and take your foot off of the brakes for a bit.

Don't shut the door in the face of these impulses. Just don't go running down the street with them.

Here's an example: there may be a coworker who really annoys you—really presses your buttons. Applying the brakes—impulse or, as Dr. Hammerness would put it, "inhibitory" control does not necessarily mean that you try to suppress those emotions completely. Neglected emotions may come back to bite you or may lead you to, say, lash out at your spouse or child inappropriately instead of the coworker.

On the other hand, don't let the impulse urging you to tell the coworker to stuff it lead you to a rash action that will undoubtedly complicate and further disorganize your life.

Listen to your rational thoughts; let the skilled rider within you pull back on those reins. Besides, you may have a chance to achieve your objectives in other ways. The right time may come to demonstrate your displeasure with these people or address the situation. Remember, as the jockey would tell you, the race is won in the long run! And it is the long run—the future—that we are looking toward here as you work to learn better control and, through it, achieve a more organized life. Meeting face to face with our impulses brings to life the creative tension between living for today and living for tomorrow.

Today is less enjoyable if we don't listen to our hearts and follow our impulses in a conscious way.

Tomorrow will be better if we invest in making it so.

CHAPTER 6

Rules of Order/*Mold Information*

FRANK ENTERED MY OFFICE FOR THE FIRST TIME accompanied by the opening chords of the rock-and-roll song "Sweet Home Alabama." Startled, I realized it wasn't his entrance music; it was his cell phone. A wiry fellow with a mustache, tattoos and a shaved head, Frank looked at me, held up one hand in a "Why me?" gesture of resignation and picked up the phone with the other. "Hello" he said gruffly, as he stood in the doorway. "Yeah, this is him, but I can't really talk…what? No, I…they didn't? Jeez, I thought I told him to…huh? Okay, okay, look, I'm sorry; we'll take care of it right away. Lemme get right back to you."

He rattled off an apology—"Really sorry, doc; gotta make a quick call"—then tapped in a number.

"Hey, it's me…about that job in Brookline…didn't I tell you to make sure we cleaned out that stuff from her driveway? She has people coming over Saturday, remember? I didn't? I could have sworn I…You sure I didn't?" He sighed heavily. "Okay, just get over there as soon you can and collect all that stuff, okay? She's pretty ticked off. Thanks."

He started to make yet another call and then clipped the phone back on his belt. "She can wait. By the time I'm outta here, it'll be taken care of anyway."

I grinned and nodded. "Nice to meet you," I said.

He sat down. "Sorry, doc," he said sheepishly. "That's the way my business is."

Frank wore patched and faded jeans, work boots and a paint-spattered T-shirt.

"Let me guess," I asked. "Are you a contractor?"

"Yup," he said, fishing out a business card. It read, "Frank's for the Memories. Home Improvements That Last a Lifetime."

"Nice card," I said.

"Thanks, doc, but there's nothing nice about my business, especially in this economy. It's a bear."

(Note: I'm paraphrasing Frank's actual choice of words here and throughout this recounting.)

"So, is it your work that brings you here?" I asked.

"No," he said, with a crooked grin. "It's my wife." He laughed. "Actually, it's something we heard on the radio the other day. They were talking about ADD or ADHD or whatever you call it. They were talking about these people who had it and how they were act-ing..." Frank now grew a bit sheepish. "And well...my wife said, 'that sounds just like you. You should get that checked out.' So here I am."

I'm not sure Frank thought he had ADHD. Many people who have the disorder have a hunch about it, but go for decades without an evaluation and diagnosis. At the behest of his wife Frank had decided to seek treatment and like many other people he finally, perhaps on impulse, came in to get a better handle on a long-term problem that was affecting his life.

"I'm glad you came in," I said to him, as his cell phone started ringing again. He clicked it off and apologized.

"It's like that all the time," Frank said. "Kind of the nature of the business. What happened here is that I was supposed to tell one of my guys to get rid of a Dumpster we'd left at a job site. That call was from the client; she was flipping out because she's having her daughter's communion party there Saturday, and the Dumpster is still there...."

"So you forget to tell him to get it out of the client's driveway?"

He nodded. "Yup."

I went out on a limb. "Frank, is this kind of thing that brought you here?"

I noticed that he slumped in his chair ever so slightly—as if the admission of this was a weight on his shoulders.

"When my wife and I heard that show on the radio, and they were describing these people as 'distracted, forgetful, fidgety'... it was like, wow, that is *me*." He chuckled as he nervously shifted in his chair. "Good thing she reminded me about the appointment with you three times... otherwise, I wouldn't have remembered it."

He was making light of it, and it was good he could do that. But there's really nothing funny about problems with memory. And in his case, it was short-term memory—or working memory. Think of it as the interaction between memory and attention. Working memory allows you to hold and process information over short periods of time and to use information as a guide to future behavior—even after the information is out of sight (something we call "representational thinking"). It's a kind of clearinghouse for the information we need to function on a day-to-day basis. When working memory isn't working well, all kinds of problems can arise. People often fluff these off as "brain freezes" or "senior moments" or describe themselves as being in a temporary "brain fog"; certainly we all forget things from time to time, particularly when we're under stress. However, I was getting the sense that Frank's lapses in working memory were not occasional. I needed to make sure.

"Tell me something. Does this kind of thing . . . you forgetting to tell your workers about moving the stuff from that client's driveway . . . does that happen a lot?"

"Oh, yeah. I probably lost a really good job a couple weeks ago because of it. It was a big kitchen and bathroom renovation. I remember finishing the initial phone call with them, running through some good ideas about the job—and then something else must have come up. Doc, not only did I forget the details of that phone call, I forgot to show up at their house when I told them I would for an estimate. I mean, I just totally blanked it out!"

"Well, do you keep an appointment book? Or take notes?"

"Yeah, I got a book. But, I dunno, it just didn't work. I used to say I kept it all in my head. Now I joke that there must be a hole up there, 'cause it keeps leaking out."

"So you used to be really on top of things? You never had this problem of being forgetful in the past?"

He sat up straight. "Doc, I've got a good memory. I remember things growing up; I remember a lot of details about things that happened when I was in high school and in the army." He stifled a laugh. "Oh man, some funny stuff I could tell you about those days. . . ."

"It's good that you can remember all those things from the past," I replied. "But we're talking about a different kind of memory here. This is remembering things that just happened in the last few days, or in the last few hours . . . or even minutes."

"Minutes?" he asked.

"Yes." I went on to describe working memory to Frank. "Have you always had problems there?"

"I'll just tell you this one time in the army; I had a late-night detail and I guess it was so dark and I was so tired I forgot where my barracks were…and I ended up going to sleep with another platoon," he said. "The funny thing was nobody even realized it until the guy next to me woke up the next morning and said, 'Who the heck are you? Would you kindly remove yourself from our barracks?'"

(Again, I'm paraphrasing Frank here.)

Interesting, I thought to myself, as he broke up laughing in his chair, in the retelling of what was obviously one of his favorite anecdotes. Frank was able to use his long-term memory to recall episodes of short-term memory loss! I was about to ask another question when he leaned close and asked me one instead. "Doc," he said in almost a whisper. "I'm only 45. Is it possible that I could have Alzheimer's?"

I'd heard this question before from young or middle-aged people who have problems with memory.

"Did anyone in your family have Alzheimer's at an early age?"

He scrunched his face up, remembering. "Nah. My grandmother was a little out of it by the time she passed, but she was in her late nineties. I think the nursing home did it to her."

"And you have noticed these kinds of memory issues, forgetfulness, since the army and since childhood?"

"Oh yeah, absolutely," Frank replied. "Drove my folks nuts. I forgot more coats then I can remember at school, at the park, friends' homes, you know."

"Okay, then it sounds more like ADHD and working-memory difficulties than dementia."

"Huh?" Frank looked confused. "Working memory? What are you talking about?"

"I thought I just told you...."

He broke up laughing again. "Just kidding. I got you going, doc."

Now it was my turn to laugh and shake my head. You couldn't help but like this guy. Still, impairment in working memory is no laughing matter. "Frank, I think we need to work on this issue," I said seriously. "I expect you could work much more effectively with an improved working memory. So why don't we..." I was interrupted by Frank holding up his index finger, and reaching for his cell phone. "Sorry again, doc," he said. "I just remembered that I got an electrician doing a job for me in Newton, and I think I forgot to give him the address."

MOLD INFORMATION

We call this next Rule of Order "Mold Information," but it really has a lot to do with working memory, the kind that Frank has a problem with.

First, let's make sure we've got our memories clear here.

When we talk about working memory, we're not talking about the memory of years gone by or the memory that recalls obscure facts or figures. We are referring to the active, working, need-it-to-function-on-a-daily-basis kind of memory—the kind of memory that can hold on to recent information and work with it, *mold* it so to speak, to allow information that is no longer right in front of you to be useful and accessible to you.

As Frank's case shows, that kind of memory—or the lack thereof—can cause all kinds of problems in your life. Indeed it may be one of the reasons you're feeling disorganized.

We'll look at that kind of memory in a second. But while working memory is a distinct form of the human ability to recall, it's useful to look at the other types—because, in general, problems with memory, in any form, are a "red flag" for many people. Start forgetting, start having the so-called "senior moments" everyone jokes about, and we begin to wonder seriously:

"Do I have dementia? Do I have Alzheimer's disease?"

A good way to distinguish the types of memory is as follows:

Short-term Memory: Who just called on the phone a moment ago; where you put your keys when you just came in the door

Recent Memory: What you had for lunch yesterday; what television show you watched last night

Long-term or Remote Memory: The name of your first-grade teacher; incidents from your childhood

As people get older, it's that "middle" type of memory that can be affected: the recent memories. This is a normal part of the aging process. But it can worsen quickly and may suggest a serious medical problem known as dementia. In dementia, memory can get much worse quickly— even over several months. With dementia, people forget about things they have done recently, including things they have done many times before, such as how to get to a friend's home or to the store. They can get lost in familiar places, become disoriented about what time it is or not recognize the people around them. They may not be able to keep track of what happens in a day.

Dementia is not simply caused by people being "stressed out" or overwhelmed by the demands put on them by jobs. Young or middle-aged adults who feel like they can't focus or that they don't know what to do next can still usually remember what they did last night or what happened at work yesterday—and certainly wouldn't frequently forget things to do. Their issue tends to be with short-term memory—as was the case with Frank. In addition, regardless of the kind of memory problem, the pattern or course of memory problems matters tremen-dously. As Frank described, his problems with memory (and some of his other issues) can be traced back to early childhood, through his days in the army and right through to the present. This is consistent with a lifelong issue, such as ADHD, and not with a recent or sudden onset of problems in later adulthood (such as with dementia).

Getting back to Frank, the problems he was describing—missing an appointment, forgetting to call someone—seemed to be related to the type of memory that keeps immediate information accessible: working memory. Let's look a little more closely at how information is processed in this form of memory.

Here's an example: The hostess of a dinner party looks at her dining room and realizes that there may not be enough space to accommodate all the guests that have been invited. As she continues on with other chores for the party, preparing the hors d'oeuvres or driving to the wine store, the image of that dining room is played back, considered from different angles and weighed with constraints of budget and time. Views of the adjacent rooms are considered, as are the menu and the guests. All of this information is weighed, evaluated and compared at the same time…and then, voilà, the solution of serving the dinner in buffet style instead of sit down is arrived at. Most of this thinking was done without looking at the dining room itself.

Let's stop for a moment and appreciate this working-memory ability: the brain's ability to hold streams of information, analyze them, process them and use all of this information to guide a future action is remarkable and necessary in order to be organized. Imagine the experience of Frank faced with the task above. Okay, maybe a fancy dinner party is not his thing, but he wants to try it out. However, just as he is driving away from the house, contemplating seating solutions, "Sweet Home Alabama" blares at him, and…it's over. He pulls over, takes the phone call, attends to whatever the crisis is on the job and completely forgets about the seating arrangement. That information—the layout of his dining room, who's coming, what he had been thinking about where to seat them—is all gone and unavailable. He must head back to the house to start over.

You can consider this brain skill as reflective, not gut-reacting, seat-of-the-pants thinking—as valuable as that can be in certain cases (and we should note that someone like Frank may be very good at that kind of thinking—what to do when one of his workers calls and says he just spilled a can of paint all over the client's floor, for example).

As a management theorist once put it, the brain that is good at molding information and representational thinking is the "tomorrow" mind, as opposed to the "yesterday" mind. That doesn't mean that someone with good working memory can't think about the past or that someone without it can't function in the present. What it means is that the mind that is adept at this Rule of Order is the mind that takes information, steps back, considers and reflects—often looking at things in new and different ways. The ability to mold information is a problem-solving step, as well as an analytical and creative step. And here again it can apply to anything, not just business situations, and it is so critical in the process of being organized. Although, chances are, some people are more comfortable molding the information visually, verbally or spatially, this is a skill to know, embrace and develop.

THE SCIENCE OF WORKING MEMORY

Because working memory is considered to have such a key role in how we think and act, there is a wealth of scientific study on the topic. The mental workspace, as working memory is called, can be seriously disrupted in different forms of emotional and neurological disorders and diseases, and there is a range within that that is considered "normal" as well. Researchers are actively debating and studying the topic—trying to decide how best to study it, how best to describe it, and what the fundamental limits of working memory are.

Here's the latest thinking on this kind of thinking: while it has been suggested that three to four "items" (thoughts, impressions, facts) is the limit of typical working memory, it is also possible that there is no set number, that instead there is a flexible range that depends on the memories being loaded—e.g., the more complex the information, the fewer total items. In a recent study published in the journal *Science,* authors from the Institute of Cognitive Neuroscience, University College, London, concluded that there are highly flexible limits to information capacity stored in working memory. The researchers described the process of working memory this way: as you direct your attention to something (as seen by the movement of the subject's eyes), you dedicate more memory resources to it than other things, so it is remembered in more detail. But this detail can fade, as you move on to focus on the next piece of information. It's as if your brain is like one of those old "instant" cameras. Long before the days of digital imaging, these were cameras that printed out a picture, and everyone huddled around it excitedly, watching it "develop" in front of you, the image gradually coming into focus. Think about that happening as you focus in on something—the picture of what you're seeing takes shape and becomes crystal clear—but now, when you turn away, imagine the picture fades back out, as you move on to the next "scene," the next piece of information. Perhaps some people can hold on to precise details for longer periods of time, but everyone has a limit.

Brain wave and neuroimaging studies have examined working memory in order to test its limits and see which brain areas are involved. These studies have shown increased activity in specific brain areas as an individual "flexes" his or her working memory, perhaps representing the firing of large groupings of nerve cells, active at the same time but slightly out of step with each other so that different information is actively held. The areas involved include the frontal and parietal lobes and the horseshoe-shaped hippocampus, located within the temporal lobe near the amygdala, a key structure for learning and the development of long-term memory.

In these studies, researchers have shown evidence of an interaction between attentional and memory brain systems—meaning that your attention and memory networks are working closely together to produce this remarkable and valuable skill of working memory. In a study of this interplay between attention and memory (with the provocative title "Conducting the Train of Thought"), researchers described working memory as limited by the ability to sustain attention to a task.

Now, let's put all this in perspective, in terms of the bigger picture of this book.

Remember, we are building a more organized brain. After taming our frenzy and achieving sustained focus, we can now mold information as we hold multiple streams of information in our working memory. Things are starting to come together nicely on our quest to become better organized.

In following these first three Rules of Order, our brain networks are truly active, engaged and firing away. And it takes networks—brain networks working together—to get to this stage of the game and this level of cognitive complexity. Specifically, with molding information, we are talking about the interface (or networking) between attention and memory. We want to support these networks; we want to improve these abilities, knowing that, like all things, our abilities can decline with age. But the good news is that brain function *can* improve. Before we turn it over to Coach Meg to find out how we can make those improvements, let's close with one last example of the remarkable modern science in working memory.

A group of researchers from Japan recently discovered evidence of a "dose-dependent positive effect" (researcher talk for more memory training with greater effect) on the integrity of white-matter brain connections involved in working memory. White matter is the connections of brain cells, the communication highways in the brain. This finding supports theories about the plasticity of the brain—that the brain is plastic, flexible and mutable as opposed to hard and inviolate; this means that skill building can yield brain changes for the better.

That's encouraging news for individuals like you who want to improve the quality of their thinking and the quality and capacity of their organizational abilities. The organized brain we've been discussing is indeed within your power to create.

COACH MEG'S TIPS

Recently, we heard a commercial speaking to the issue of working memory—in a very different way.

Do you walk into a room and forget what you went in for?
Do you forget the names of people you know well?
If so, we can help end that 'brain fog' today!
Yes, improved mental clarity and focus is just a phone call away!

This commercial wasn't for a research project but rather for a product—and a questionable one at that: a drug, available by mail order, that allegedly sharpens memory and clears away this so-called "brain fog."

The truth is that there is no magic bullet for memory improvement. While there are pharmacological treatments that can help in some clinical cases of ADHD—such as in Frank—these are not the same as non–FDA approved, mail-order drugs, the kind being touted in commercials like these. We certainly don't recommend them.

However, despite some encouraging findings like the study in Japan that Dr. Hammerness referred to, it's worth noting that the whole question of whether working memory can be improved is still contentious. "There is no penicillin for memory," says Dr. John Hart, a neurologist at the University of Texas's Center for BrainHealth in Dallas. "It's clear that pharmacologically, you can change someone's working memory abilities. There have been other studies and techniques that people have tried and are using. But I have not seen one that has worked 100 percent."

Neither have I. But there are experts in working memory and health professionals on the frontline who have found techniques that are effective at least most of the time and that I believe can help you as you develop your skills in this Rule of Order—the ability to mold information and to use your working memory—which is a vital step in our journey toward a more sane, better organized life.

But first, just as Dr. Hammerness did earlier in this chapter, we need to step back and remind ourselves of the big picture here: at this point in our pursuit of a better ordered, saner life, we've tamed the frenzy, our attention is focused, and we've learned to control our impulses and apply the cognitive brakes when necessary.

Now it's time to really get organized.

Our last three rules are where the rubber hits the road, where we plug in all of the channels of our working memory and leap to new insights. From there, we can connect the dots and see the bigger picture—whether it relates to losing keys, having a difficult conversation with a work colleague or figuring out where we'd like to be ten years from now—which is hard to do when you can't seem to get through today without losing something or getting distracted.

One of the great things about being a coach is that I get to go through this process every coaching session; the intense collaboration of two brains produces an expansive working memory, sparks plenty of insights and delivers a fast path to higher ground—where one can look down on the frenzy left behind. When doing this, I often look for metaphors to help explain some of our change strategies. The one I use for working memory comes from my love of music.

As an audiophile, I imagine working memory as multichannel stereo speakers, a jazz band or an orchestra. Lots of channels or voices all plugged in at the same time, layered one on top of each other and working together to make an integrated whole. I sometimes stretch out my hands and fingers before a coaching session as if to switch on ten channels of my working memory and call them to the task at hand. I can feel my eyes and my mind grow wider as I stretch to hold all of the channels in the moment. It's pure pleasure to draw on each channel of working memory, one after another, when working on a task. You can harmonize these memories or shine a spotlight on one channel—or, to keep to our musical metaphor, turn up the dial and listen as this memory sings to you, sometimes loudly, other times more faintly. Like isolating the string section in a symphony or the bass line of a great jazz song, you can tune into it. Take a bit of memory and turn it over and side to side in your mind, savoring what it has to offer before moving the spotlight to another bit of memory. Amazingly when you shine the light on a bit of memory and it's gone—ugh! It's gone, can't remember it!—you can move on without fretting, and that forgotten nugget of information will pop up later.

I offer that to you as a way to look at working memory—not as *work,* as its name implies, and not as an obstacle, as it often seems when we can't remember a specific fact or name at the moment, but rather as a process or maybe even a pleasure.

Let's look now at ways to help sharpen your working memory—one of the key steps on the way to better organization.

For our tips on how to improve the ability to mold information, we depart slightly from the approach in the last few chapters. I'm drawing on some tried-and-true advice from professionals in the "memory game," in particular Dr. Marie Pasinski, MD, a neurologist at Harvard Medical School and author of *Beautiful Brain, Beautiful You*, and Martha Wolf, Director of the Alzheimer Center at Parker Jewish Institute for Health Care and Rehabilitation in New Hyde Park, New York, one of the country's foremost treatment centers for Alzheimer's patients. Martha is a frontline professional in a field where memory loss is not just an inconvenience but a critical function.

Some of their memory-sharpening techniques are evidence-based; others are common sense or tested in the crucible of a rehabilitation center where the anguish of age-related dementia gives us all pause to put our own organizational challenges in perspective.

Either way, these can help us sharpen and improve our ability to remember and to mold information more effectively.

Sleep to rest . . . and remember

You've heard me stress the importance of sleep. But according to Dr. Pasinski, it's particularly important for this Rule of Order. "The role of sleep in memory consolidation is huge," she says. "It's during sleep that we actually process new information."

How much sleep? The general recommendation is seven to eight hours. "Some people can get by on six hours a night, but most people think they can get by on less than they need," she says, adding that *quality* of sleep is as big a consideration as quantity. "I think what's important is that you wake up and feel rested. You should be able to wake up in the morning and feel refreshed, not be dragging or dependent on caffeine to get you through the day."

One other tip for restorative sleep, Pasinski says, is to keep regular sleep and wake cycles. When our circadian rhythms are disrupted, hormonal levels and neurotransmitter functioning in the brain can all be affected. This, in turn, will compromise our brain function, and our ability to use information. "A regular sleep schedule is what your brain needs," she says. "If you're not getting adequate and regular sleep, you're not going to remember things as well."

Consider time when developing memory

When you choose to learn important new material and information will affect how well you remember it and how available it will be when you need it.

Here again, sleep plays a role. Pasinski cites a study in which subjects were far more likely to learn a new skill when they were taught the night before. Following a good night's sleep, they showed greater improvement than if they were taught the skill during the day and tested that night. "They really did sleep on it!" Pasinski said. "The study suggests that we actually learn when we sleep."

Because learning is reinforced during sleep, it's a good idea to get a good night's sleep after learning the new material. You will likely perform better than if you didn't get some shuteye in between the time you learned or practiced the skills and were tested on it.

Also, when you prepare, break your study sessions down to two sessions, so that you can cover the same ground twice. "Repetition is really important for learning," Pasinski says. If you have an hour to practice something you're better off practicing in two thirty-minute sessions...covering the same ground twice...which seems to reinforce learning.

Flex your memory muscles

In *AARP The Magazine* articles about how to help delay the onset of Alzheimer's or senile dementia, you'll hear about the importance of keeping your brain active by doing puzzles. While Pasinski agrees with the importance of training your brain, she suggests other ways to give your memory a mental workout. "I think a better use of your time is to learn something new," she says. "That's a wonderful way to challenge your brain, it's fun, and gives you a sense of accomplishment."

If you're retired or someone with time on your hands, maybe French lessons, stamp collecting or learning how to play the guitar would be an excellent investment of energy. If you're reading this book, though, chances are you don't have a lot of time for those kinds of pastimes or hobbies, rewarding as they may be. I suggest you make your mind-stimulating, learning exercises practical.

One way to do this is to direct your reading toward something you don't normally look at. Instead of, say, *Time* or *Fortune,* read *The Economist* and get a global view of politics and the economy. Instead of *Sports Illustrated* or *Woman's Day,* check out *Wired* or *Salon.* Instead of *The New York Times,* read *The Wall Street Journal.*

This tip isn't limited to newspapers and magazines. If you spend a good deal of time in the car, purposive listening to books on tape is an excellent way to sharpen memory. Wolf suggests structuring it for maximum memory-building benefits. "Listen to one chapter or section a day," she says. "Then before you listen to it on the drive home or the next day, make a point to summarize in your mind exactly where you left off . . . what's happened in the plot." That, she notes, will not only enhance your enjoyment of the book but also make it more valuable as memory training (and of course, if you don't spend a lot of time in the car and instead read your books the conventional way—in print or on a digital reader—you can do the same thing).

The key point in all this reading and media cross-training: "You're challenging your brain by learning, by making new connections," says Pasinski.

And there's a bonus here: No matter what your industry or profession, it never hurts to expand your perspectives and broaden your horizons. You never know where the next great idea or nugget of information will come from that can help you in your career—and the fact that you'll be more likely to remember that idea or nugget because you've been training your memory muscles makes it even more worthwhile!

Create a memory book

Throughout this book, I've talked about the importance of writing things down: your goals, your visions, your observations on your own behavior. Martha Wolf has another kind of writing assignment for you that could be very beneficial. Here's how she explains it:

"Sometimes we can improve our memory, sometimes we just need a crutch. That's why I talk about things like putting the items you need in a place where you most need them. We get crazed because we can't remember where we put something. So don't make yourself crazy!"

To help keep track of these items, Wolf suggests creating a memory book. This is a tool that she has used with the families of her patients, who very often find themselves facing a situation where a parent is incapacitated and they, the children, have no clue where any of their parents' important papers, accounts or keys are located. But the idea can be adapted for those who are having trouble with more mundane household items of their own.

"It's a good thing for everybody to do," she says. "And it takes a lot of pressure off."

Wolf suggests an old-fashioned composition book for her patient families, but you can create a document online as well. "Make a list of the things you use most frequently—extra sets of keys, eyeglasses, wallet—and find the most logical spot for them. So, for example, if you read before you go to bed, you would probably put the eyeglasses on your nightstand. Then write it down in the memory book."

Argue to improve working memory

Wolf frequently speaks to senior citizen groups. Inevitably, one of the first questions she is asked is how to keep mentally sharp. "I used to tell them, 'Go start an argument,'" she says with a laugh. "Then I realized that didn't sound very nice so I modified it."

Her point is that arguing—not screaming and yelling but reasoned debate—is one of the best mental exercises you can do. "When you have a difference of opinion, you're listening intently to information, and while you're listening you're formulating your response," she says. "You have to be mentally agile and nimble. It's like a game of cognitive ping-pong."

Short of joining your local debating society, a good way to get the mental benefits of arguing is to watch one of the all-news cable channels where the polarization of the American political scene may actually have a salutary effect. "If you normally watch CNN, turn to Fox," Wolf says. "If you watch Fox, turn to CNN. Listen to what their commentators are saying, develop your response, put the television on mute and talk back to them. Yes, it's okay to talk back to your television…just make sure you let anyone else in your house know beforehand so they don't think you've gone crazy."

Talk with your hands

One of the more unusual but intriguing tips for improving memory comes from another Harvard colleague, psychologist Jeff Brown, co-author of *The Winner's Brain*. "Gesturing in a meaningful way while you are learning may help you when recalling the concept," Dr. Brown says. "The idea is that you are storing at least two different types of information about something you'll need to recall later. A good example of this is when kids speak math problems aloud but also 'work them' in the air." (We know some adults who still perform arithmetic that way, so it's not just children who rely on this method!)

Brown suggests that when you've just learned someone's name, "write" it down on the palm of your hand with your finger. The act of tracing the letters on your palm can help your brain remember it, says Dr. Brown. Or, he says, "air-write on an imaginary map of your grocery store or mall as you name aloud the items or stores you need to remember when shopping." (As you would when you're arguing with your television, make sure you perform this memory-enhancing drill discreetly!)

Remember to exercise, and exercise will help you to remember

I've been touting the benefits of regular exercise throughout this book. If we haven't convinced you yet of the mental benefits of physical activity, here's one more piece of evidence.

A 2009 study published in the journal *Hippocampus,* found a positive relationship in older adults between physical fitness and the size of the hippocampus—a key brain structure thought to be the central processing area for memory and learning. Unlike some other parts of the brain, neuroscientists say it is "plastic"—malleable and dynamic. Although the hippocampus begins to degrade as you get older (its volume shrinks about 1 percent a year after age fifty-five), it can respond to positive stimuli, suggesting that it is a "use it or lose it" organ. In the study, neuroscientist Kirk Erickson of the University of Pittsburgh wanted to see if physical exercise might have a positive effect on the hippocampus. His team tested the fitness levels of 165 adults who were over age fifty-five and also gave them brain scans and spatial memory tests. The findings: "The fitter subjects had hippocampuses that were about 35 to 40 percent greater in mass than sedentary individuals," Erickson says. The result surprised him. "I wasn't expecting that big a difference," he says.

How much exercise do you need to get the brain benefits? Although the fitness levels of the subjects ranged from sedentary to moderately fit, Erickson says, "None were super athletes." In other words, you don't need to run marathons to keep up brain size, just regular exercise.

This echoes similar findings in other recent research: A 2008 study in the journal *Neurology* found that seniors who regularly took walks had a lower risk of developing vascular dementia, a kind of memory loss associated with inadequate blood flow to the brain (it's the second most common form of dementia, behind Alzheimer's). The study, which was conducted in Italy, followed 750 older men and women over four years and found that those who were the most active—the top third— were 27 percent less likely to develop vascular dementia than those who walked the least.

The results, notes the website Alzinfo.org, are consistent with other studies on the relationship between moderate physical activity and brain health, including a 2004 study of more than 2,200 older men living in Hawaii, which found that those who walked the least—less than one quarter mile per day—had nearly twice the risk for developing Alzheimer's and other forms of dementia than men who walked more than two miles a day. That same year, the even larger Nurses' Health Study from Harvard reported that women in their seventies who engaged in regular physical activity like walking did better on memory tests than women who were less active.

The findings of all of these studies are indeed encouraging for older adults, but it's important to note that the value of walking also applies to younger minds and memories with far fewer years to reflect upon. "Getting your heart going and increasing the blood flow to your brain with a twenty-minute walk can have very positive effects on your brain," says Pasinski.

This prompts her to offer a suggestion—which is healthier and a lot cheaper than the questionable one on that radio commercial I heard: "When you're in a so-called 'brain fog,'" says Pasinski, "get up and get out. The exercise, and just the change of scenery, will get you out of that feeling of mental staleness and help make your memory sharper."

CHAPTER 7

Rules of Order / *Shift Sets*

N ICK WANTED TO BE AN EMT, and at first glance, he appeared as if he was well suited for the job. Certainly, he seemed able to meet the basic requirements for that job, as specified by the Massachusetts Office of Emergency Medical Services.

He was over 18. (Nick was 23 when I first met him.)

He could speak English. (Nick grew up in the Greater Boston area; he could also speak Spanish.)

He could lift 125 pounds. (Or at least I'd be willing to bet he could. When we shook hands on first meeting, I thought my fingers had been put in a vise.)

His desire to be an EMT was laudable. I got the sense in just a few minutes talking with him that he was driven by a genuine desire to help people and save lives. First, however, he had to work out a problem. The state certification course consisted of thirty-three lessons, involving one hundred hours of classroom and field training, plus ten hours of in-hospital observation and training. Like every other aspiring paramedic, Nick was going to have to study hard. And while I could tell he was determined to do well, he was already falling behind and feeling overwhelmed. That's what had brought him to my office.

"Tell me about it," I said. "What's going on in the classes?"

"It's very interesting—what we're learning," he said, "but I'm having trouble." He told me what had happened during a recent lecture. The topic was how to respond to someone who might be having a stroke. First, the instructor needed to explain what a stroke was, how and why it occurs, symptoms, risk factors and so forth.

"There was a lot of information, so of course I took notes," Nick said.

"Of course," I agreed.

As Nick went on to describe it, the notes he took were voluminous and involved. He even started drawing pictures and diagrams of the brain, based on what the instructor was saying about the causes of strokes. About halfway through the lecture, however, he realized he was the only person taking notes.

"What was everyone else in the classroom doing?" I asked.

"They were watching the instructor demonstrate on a dummy," he said. "He was showing the class the best way to transport a stroke victim." He had also used a teaching assistant as a "victim" to go through the standard questions and assessments that EMTs ask in order to determine whether a stroke has actually occurred and its extent. But while the instructor had stopped lecturing and started demonstrating, Nick was still busily making notes.

"Did you just not realize he had stopped lecturing?" I asked.

"It's hard to say," Nick said. "I guess, sort of. But I just couldn't pull myself away from the stuff I was writing. I needed to finish that."

Nick did say he was embarrassed when he noticed some of the other students in the class giving him sideways glances as he continued scribbling away, while everyone else had put down their pens. I also got the sense that it wasn't the first time. "This kind of stuff has happened before," Nick admitted. "I've never done well in school. For some reason, I'm not good in the classroom." He attributed his subpar scholarship to deficiencies in his own intelligence or abilities. "I guess I just don't have what it takes to succeed in school," he said. "I've always had trouble learning. I just can't keep up."

I suspected that the problem was really not that vague nor that it had anything to do with intelligence. Nick struck me as a bright young man, and based on what he'd told me, he obviously could focus and take notes. But we needed to learn a little more about this apparent inability to shift gears when everyone else in the class had.

"Does anything like that ever happen at home or in your social life, outside the classroom? Do you ever feel sort of set in your ways or unable to change or be flexible about things?"

"You mean like 'rigid'?" he said, with a grin. "That's what my old girlfriend called me. She said that I was never flexible, couldn't be spontaneous. That's one of the reasons we broke up."

He went on to relate an example. "Thursday night was our movie night when we were dating," he said. "Then one Thursday morning, she called me and told me that someone from work had two tickets for that night's Sox game. We're both big Sox fans, so she was really excited about going."

"What did you do?"

"Part of me wanted me to go," Nick said. "But I was like, 'tonight's movie night.' And we had this whole routine. I'd go to the gym after work, we'd meet for dinner and then catch a movie. That's what

I was expecting we'd do. I mean, I really wanted to be spontaneous, but I just…couldn't. That's hard for me."

There were other examples. He told me that he'd been having trouble in the gym as well. "I've been doing the same workout for years," he says. "This friend of mine, he's a trainer, told me it would be good for me to get more core strength as an EMT since I'll be lifting stuff around, and your legs and your core muscles are really important for that. He sent me this routine that he's used, it was kinda cool, using medicine balls and other stuff."

The friend even dropped by Nick's gym to show him how to do these new exercises. But when his buddy arrived, Nick had already started his regular workout—his bench presses, shoulder presses and biceps curls—and he just couldn't bring himself to stop. "I kept saying, 'let me just finish one more set,'" Nick said. "Finally, my friend just shook his head and did the new workout on his own."

In the field of psychiatry we define impairment as something that gets in the way, manifesting itself across multiple situations, not just one setting. Simply deciding not to take advantage of the baseball tickets because Thursday was movie night is not by itself evidence of a problem (although I must admit, free Red Sox tickets is not something to be taken lightly in this town!). Still, when it creeps into every aspect of life—in Nick's case, into the classroom, the gym, as well as his personal life—and when he feels stuck with this mind-set, despite evidence that it's causing him problems, then it becomes more serious. Nick's problem was not that he was "just set in his ways," if only because, in my opinion, he was a little too young to have set ways. Nick's problem was that he was not able to shift his attentional or behavioral set. He could not easily change his focus from one thing, situation or setting to another.

SHIFTING GEARS

Before we look at what Nick's case can tell us about the next Rule of Order, let's step back a moment. By getting to this point in the book, and following some of the suggestions that have been made in the previous chapters, you should have a solid foundation for a more organized and less stressful life. If you've been working on integrating the first four Rules of Order, you should by now be able to

- approach a task more calmly, as you have (to use our terminology) "tamed the frenzy" (or if not tamed, at least held it at bay at key moments!)
- sustain your focus
- put on the cognitive brakes when it's necessary
- use your working memory to mold information

This doesn't mean you'll never be flustered or get distracted again. But now at least you have the beginnings of an operating manual so to speak, a plan that can help you start facing the day with greater confidence.

So let's continue to build on these successes.

But as we do that, it's important to remember this: although we have broken these various cognitive skills up into discrete rules or steps, in reality, they are closely interrelated. As we have learned, brain regions tend to work together, not independently. It's less like a weight lifter doing curls that isolate the biceps muscles of the arm and more like a batter using his or her legs, trunk, abdominals and biceps in tandem and in coordinated fashion, in order to swing a bat and drive the ball over the fence.

Such is the case with this next Rule of Order—what we call Shift Set. This refers to the ability to be flexible in your thoughts and behaviors.

In order to be organized, you must be able to effectively and efficiently shift your focus or "set" from one object, action or situation to another. In so doing, you can move on to the next action—take the off-ramp if it leads to a better road or a new opportunity. Without this skill, our attentional system starts looking more like tunnel vision (as was the case for Nick that day in the classroom). We need to be nimble and ready to adjust and to shift our focus and our behaviors when necessary, in both our work and personal lives.

What does it mean in practical terms?

Successful set shifting is you being able to pull yourself off of the interesting news article you were immersed in in order to answer the emergency call from a colleague. She needs some details on the big project you're working on before going into a client meeting, and she needs them now.

Successful set shifting is you at a meeting where you were prepared to present some important data. But when others in the meeting steer its focus on to some other topics, you don't get flustered or annoyed that you're not getting your turn and that the work you planned to present is not going to be heard. Instead, you switch gears to engage and participate in this new discussion.

Successful set shifting is you coaching your kid's basketball team and changing your practice plan because only three kids showed up.

Some people do this naturally. Others (like Nick) have a very difficult time making these shifts. And if you can't—can't switch gears to pull together the information your colleague needs, can't go with the new flow of the meeting or can't modify practice—you're going to find yourself frustrated, overwhelmed and disorganized.

In these examples, perhaps you can also see how set shifting may be inextricably tied to a couple of our previous steps.

How can you shift directions if you haven't first successfully applied the brakes?

And how can you shift with confidence if you aren't using your working memory? As you are shifting your attention, you are leaving a path behind you, one which you may want to hang on to, to remember in your working memory and to guide your way. You should shift with a sense of intent or *planning*. Now that shouts "Organization!"

SCIENCE OF SET SHIFTING

As we have seen before, scientists typically study cognitive processes like set shifting by having experimental subjects perform various kinds of tasks. Thanks to brain imaging, researchers are then able to observe the subjects as they perform these tasks and to see which parts of the brain "light up"—or are activated.

The original test for what we call set shifting, however, far predates the advent of modern brain-imaging technology. One of the oldest and still most reliable tests of cognitive agility, the Wisconsin Card Sorting Test (WCST) was developed at the University of Wisconsin in the 1940s by a team that included the famous American psychologist Harry Harlow (who also conducted groundbreaking primate research on the impact of love and affection in development). The test was first described in a 1948 journal article titled "A Simple Objective Technique for Measuring Flexibility in Thinking." In the WCST, subjects are asked to sort and stack a series of 132 cards. But the rules of how they are to be stacked change, unpredictably, during the course of the test. The degree to which the subjects can adapt to the new rules is a measure of their mental flexibility—or what we would now call their ability to set shift. It has been found that those with brain (frontal and prefrontal cortex) damage will get stuck in one sorting modality. ADHD sufferers often have a similar problem—an inability to easily shift or adapt to the new rules presented.

While the WCST has had an important role in our understanding of and ability to measure set shifting, current investigations are attempting to isolate the very specific brain regions underlying set shifting in order to understand better how this process works.

In a 2010 study, scientists at Stanford and the Michigan Institute of Technology investigated the brain regions shared between the skills of inhibition (applying the brakes) and set shifting and tried to determine what brain areas are unique to each. In this study, healthy college students were presented with a series of large letters made up of smaller letters in various colors. Subjects were asked to identify large letters or small letters as cued by color. Sometimes the large and small letters were the same (dozens of little *h*'s that made up one big letter *H*) or different (little letter *s*'s to form a big letter *H*). Sometimes subjects were asked to toggle their focus back and forth between the small- or large-letter components—in other words, shift sets—and sometimes they were asked to simply focus on the large or small letters.

Through brain imaging, the investigators were able to watch which parts of the brain were active as the subjects performed the tasks. They found that one specific brain cortical region was particularly important in the set-shifting process: the inferior parietal cortex, which we talked about earlier in regard to its role in helping us both sustain attention and consolidate memories. But—true to the brain's cooperative nature—the parietal cortex did not work alone. It seems that a network of brain regions (prefrontal, parietal cortex, basal ganglia) all work together during both inhibition and shifting, suggesting that the two are actually one major process. Or perhaps that inhibition is a needed component or a sort of "prerequisite" for the more complex task of shifting. (That's how we introduced these concepts in the Rules of Order and that makes intuitive sense.)

Given the complexity of this task and the number of brain areas involved, it again makes sense that this ability to shift sets appears to be a skill we develop as we mature. For children, transitioning from one activity to another is often a painful process (and, as any mom or dad can attest to, it's no picnic for their parents either). Of course, there are exceptions such as Nick, who continued into young adulthood to have difficulties making transitions and shifting sets.

As we get older, shifting again becomes more difficult (something that anyone dealing with aging parents will recognize). This is often blamed on older adults being "stuck in their ways" or intransient and unwilling to change. A group of researchers at the University of California, San Diego recently demonstrated that the decline in one's ability to set shift as we grow older may be due to a degradation of the so-called white-matter tracts that we've mentioned before, those neural "highways" that connect brain regions. The tracts are composed of whitish, insulating tissue called myelin that surround a string of nerve cells. Like the asphalt interstate highways that crisscross America, these routes link the various regions of the brain. Some of the tracts are straight and short, others long and winding. As we get older, the integrity of these roads—like the highways we drive on—can degrade. They get bumpier, there are more potholes and moving information along them becomes a slower, more difficult process. Older adults, whose brains are connected by this aging infrastructure of white-tract matter, must proceed with caution and deliberation—not because they necessarily want to but because they have to.

WHY BOTHER WITH ALL THIS SHIFT AND CONTROL? JUST MULTITASK!

Here's some advice from various online authorities about the wonders of multitasking; this is one about women.

- Any good housewife, mother, housekeeper, cook, secretary or waitress has always known about the benefits of doing several things at once, and she's a whiz at doing it!

- In order to be productive, you need to do many things at once—and effectively!

- How do you manage to stay sane when you're insanely busy? You become a master of multitasking, of course!

Well, there you have it. No need to read any further, is there? The secret is doing three or four or five or six things at once. That's the real key to organization, right?

Wrong.

Back to the automotive analogies and our last few Rules of Order.

Inhibitory control is your driving a straight and true course and applying the brakes when needed to avoid going off the road due to some distraction.

Working memory allows you to recall the road back and that side road you just passed, which may just have the only gas station in the town.

Set shifting is your being able to turn the wheel on a dime and redirect the car when that distraction is worthwhile, significant or valuable. The ability to do all three is a sign of the organized mind. What the organized mind *cannot* do is drive in three different directions at once. Yet some of us are trying to do just that. I'm talking here about that much-used and -abused term *multitasking*. A concept borrowed from the world of computers, a great deal has been written about this supposed ability to skillfully juggle many tasks at once.

In the 2006 story "The Multitasking Generation," *Time* identified the teenagers who seem to text, burn CDs and do their homework at the same time as the main perpetrators of this new and questionable approach to life. But the truth is that multitaskers come in all ages and settings. "In most project environments multitasking is a way of life," wrote management consultant Kevin Fox. "This seemingly harmless activity, often celebrated as a desirable skill, is one of the biggest culprits in late projects, long project durations and low project output."

Still, many people boast that they are good multitaskers; the implication is that they are somehow performing several tasks at once—talking to you on the phone, writing a report and . . . who knows? . . . knitting a sweater and monitoring the evening news all at the same time. While you plug along, trying to do one thing at a time, these people have left you in the dust as they speed along through a hyperefficient life, adroitly handling several tasks simultaneously. Those who see themselves left behind lament at what they perceive as their inability to do this. "I need to learn to be a better multitasker!" is a complaint I sometimes hear from patients.

No they don't. And neither do you.

There are indeed important cognitive skills you can learn in this book; and, as we said in the beginning of this chapter, if you've been reading and applying Coach Meg's prescriptive advice, you are probably starting to utilize better the innate organizational tools and abilities of your brain already. But I'm afraid that multitasking is not one of these. Despite the glowing promise of multitasking enthusiasts, the idea that you can really attend to many things at once is simply a myth of the pseudoorganized. Certainly, it would be nice if we could simultaneously work on four or five tasks. But it's simply not true. Trying to do a multiplicity of tasks well at the same time usually leads to one end—all of those tasks done inadequately or incompletely. Multitasking is your trying to stretch your ability and to do more than you can. It's the illusion of a lot of balls in the air. In a snapshot, you look like you're juggling six different things, but in the next frame, they all fall to the ground.

A recent study showed that those who identified themselves as heavy media multitaskers processed information differently than infrequent multiple-media users. Were these masters of the media universe better able to process tons of information on various channels? On the contrary. Heavy users did *not* filter out irrelevant stimuli well, they did not ignore extraneous information and they did *not* switch tasks well.

How could they? Do a little experiment for yourself: Take out your cell phone, turn on your television, and boot up your computer. Now try texting a friend, watching a television program and listening to iTunes on your laptop . . . all at once. Can you do it? Sure. But can you do it in any meaningful way? No. You will inevitably say little, see little, hear little. To those of us who study the brain and its impact on behavior, this is no surprise. We're simply not designed to attend to multiple inputs and perform many tasks at once. On the other hand, we have an innate and marvelous ability to concentrate deeply on one subject but very quickly pull our attention off of that and apply the same level of focus on something else that we have rapidly surmised is of greater immediate importance. So if you focused most of your attention on composing your text to your friend, then shifted to watch the television program and then changed sets again to concentrate on the music, the level and quality of the experience for each would be far, far higher.

That ability is what we call set shifting and that, along with its "partner" process—inhibitory control—is what you need to practice doing more, in order to truly become more productive and efficient.

When it comes to getting things done, as a twelve-year-old boy might colorfully phrase it, set shifting kicks multitasking's butt.

Let's look at a practical example of how and why that is.

MULTITASKING YOU, SET-SHIFTING YOU

Think of yourself for a moment as the typical multitasker at home with the kids on the weekend. You start off doing the laundry and then see some cleaning to do in the cellar. As you sweep behind the dryer, you spot an old toy—a charming wooden riding horse that your aunt gave the kids—lying there. The children are delighted to see it, having long since assumed it was lost forever. A project is undertaken to begin repainting it, while the laundry is left half-folded.

Meanwhile, the morning is drawing late. Now it's time to head to the store to pick up lunch and, in returning home, to begin a project in the garden that you remembered as you were pulling into the driveway—while still partially juggling laundry and the toy-painting project. Now you've got laundry, cleaning, toy painting, gardening and not to mention lunch on your plate. You are feeling confident in your multitasking.

But, not for long. Suddenly, it's 4:00 pm, and a friend calls to remind you that you were supposed to be dropping by for a barbecue—and, oh, hadn't you volunteered to bring the fruit salad? Not to worry, you say confidently, pointing with pride to the fact that you are "multitasking." You start to list the various projects going on, and as you do you begin to realize that *none* of them have really been completed; *all* have resulted in more work. Most of the laundry has been washed, but none ever made it to the dryer (the discovery of the toy sidetracked you), and now some of it will have to be redone. The cellar is still a mess. That toy is still unpainted. The cold cuts that you left out on the counter that were supposed to be made into sandwiches have spoiled in the heat. There are now half-dug holes in your garden; the implements of which are lying about and now have to be put back in the garage.

Oh, yes, mighty Master of Multitasking, you've really got a lot of things going on this Saturday! The problem is that they are all unfinished, resulting in more work for yourself and greater disorganization around your house. Oh, and by the way—you arrived late for the barbecue, and by then everyone had eaten and no one was really in the mood for your fruit salad.

We are not aiming to achieve robotlike efficiency or effectiveness every living minute of the day. But on the other hand, a multitasking scenario like this is inevitably discouraging (trust me, I've heard it recounted angrily by many an ersatz "multitasker")—and understandably so: it's very frustrating to feel that you are rarely achieving what you set out to do, and in the big picture, this can be quite damaging to your goal of being better organized.

To get back to our Rules of Order, it's time to shift from the multitasking myth to the science of set shifting and to take another step toward a more organized life. Now, let's take a second look at the scenario we just described.

Instead of you as the typical multitasker at home with the kids on the weekend, let's see how things could go differently if we made you an ace set shifter.

So instead of trying to do a number of things at once, you are now more cognitively nimble, ready to respond to opportunities as they present themselves—thanks to your good set of brakes and your skill in holding and molding information.

We witness the same beginning to the day. The laundry is brought down to the cellar and you spot an old toy lying behind the dryer—a charming wooden riding horse that your aunt gave the kids. The children have wondered where this is and are interested in reclaiming it. But this time, instead of springing willy-nilly into a reclamation project, you apply the brakes. You imagine the day's schedule on a virtual whiteboard in front of you, and then you set shift. It's 9:00 am. The to-do list for today is reviewed and altered. Besides the laundry, there is general clean-up around the house, food shopping and the barbecue later this afternoon. (Notice that leading up to your success as a multitasker is your ability to stay calm, keep your focus, apply the brakes and run through scenarios in your working memory. Sound familiar?)

You decide to complete only one full load of laundry this morning, and shift for just a few minutes to getting a space in the cellar ready for the kids to clean and paint the toy later this afternoon—after shopping, while you are prepping for the barbecue. This occupies all your attention for a short while. When it's done, you stop again, shift away from this task—without engaging in anything else—and head out to the store. When you return, you hear the weather forecast as you're putting away the groceries. They're calling for rain tomorrow (in the multitasking version of this story, you had the radio on but were too busy thinking about the other things you had left undone to really pay attention). Hearing the forecast, you recall that tomorrow, you were planning to spend some time in the garden. You realize that the garden project is probably better suited for the sunshine of today and the indoor toy painting for tomorrow. So instead of heading down to the basement, you stop, *shift* and devote a couple of hours to the garden.

There is a much more organized feel to this day now, isn't there? Maybe it seems subtle, but the payoff even in this domestic example is significant. Instead of limp, soggy laundry, a messy basement and a garden in disarray, not to mention spoiled luncheon meats and an unmade or unloved fruit salad, you have several tasks done or on their way to completion and others have been rescheduled for tomorrow.

Notice how application of brakes, retaining and molding information and shifting directions are done in a thoughtful way. In so doing, smaller tasks are defined, prioritized and accomplished. Your shifts take place in the face of new contingencies or information (such as the weather forecast). They are not rapid fire and random; they are deliberate and purposive. Here we see also how applying the brakes, molding information and set shifting are tightly related concepts. Notice, as well, the potential influence of some of the other Rules of Order on your ability to shift sets. If after hearing the forecast and realizing you couldn't work in the garden tomorrow, you worried yourself into a tizzy or threw a fit, your ability to recognize and react—to set shift—could, like your half-done laundry, be hampered.

How do we attain the nimbleness of the set shifter? As with the other Rules of Order, for most of us it's an innate skill but one that can be improved.

COACH MEG'S TIPS

My vision is that all of us working on organizing our brains become master set shifters—not just out of the need to respond to the many valid interruptions of our focused attention. No, it's more than that. The master shifter has expanded his field of vision to welcome the new opportunities that life presents us. The shifter's mental flexibility allows him to be nimble and agile, changing the "set"—or situation—fluidly. Instead of reluctantly letting go of the comfortable feeling of his focus on a task, the shifter appreciates the gift of change and what it might bring both to the new situation and to the original task as well.

Dr. Hammerness has described set shifting in terms of cognitive flexibility—which it indeed is. But it's also a skill of cognitive *creativity.*

And spontaneity—because set shifting is not always deliberate. It could happen during an interruption.

Here's an example of what I mean by the master set shifter's attitude and skill on the mental throttle:

You're in the middle of preparing dinner, and you get interrupted by a phone call. The less-than-organized mind is not prepared for this interruption, not willing to allow it much less embrace it. It's an annoyance, a distraction. The organized mind—the master set shifter—sees it differently. The organized mind stops in midpreparation and takes the call. Ten minutes later, when you resume your cooking, you remember, "Oh gosh, I have fresh basil growing in my garden! That would really add a nice taste to this pasta sauce."

Voilà! You unhooked the brain from its tether and allowed it to wander off from thoughts of recipes and cooking during the phone call, and in that space, new ideas bubbled up, as welcome as the warm and delicious aroma that is emanating from your pot.

Skilled set shifting can be strategic as well. Think about how it could help you at work. You're plugging away at that PowerPoint presentation you have to deliver, and you arrive at this point of diminishing returns where you just feel you're spinning your wheels. You could throw your hands up in disgust, adding to your feeling of disorganization and disarray. Instead, a strategic set, stop and shift can change the game— when you say "I'm not getting further on *this,* so I'm going to jump on to *that.*" Make a phone call, check e-mails, take a break. That's the deliberate intention choice—and very often it will result in the fresh idea and the new perspective that pops into your mind when you jump back on the original PowerPoint task.

We chose that verb—*jump*—deliberately. A shift is like a leap, letting your feet leave the ground for a moment and jumping from one wire in your brain to another. The shift brings new insights, new ideas and new thoughts that will eventually improve your performance on the task temporarily left behind. A deliberate shift is an opportunity not to be missed to get out of a trench and rise to a higher viewpoint.

You can't force set shifting—it's almost a state of mind; when you suddenly remember the basil that makes the meal, you have no idea why talking to your mother on the phone brought this on. But it did.

Still, while you can't summon up a set shift on command, you can create the conditions for it to occur when the opportunities present themselves. You need to be open to it and ready to make that leap.

Here are some suggestions on how to make yourself the master set shifter—a critical step in getting yourself better organized.

Get light on your cognitive feet

Except for professional basketball players or track stars, most of us find it hard to jump, and don't do it very often unless we are asked by a cute leader in spandex of an exercise session. Gravity is a tough force to beat. Sometimes we are prone to resist jumping because it feels more comfortable to stay in the groove of a task.

If you've ever woken up suddenly from an intense dream, you'll recall the sense of amazement at your brain's ability to leap around in wild ways to wild places. Unbridled by the demands of the day, our brains show their raw potential to roam untethered to reality.

Let's wake up and put the "roaming" power of our brains to work. Remember, also, that when our frenzy has been tamed and our brains are more organized, our thoughts start to feel lighter and more agile. Getting better at all of the rules that we've shared so far will help you lighten up.

Bust a silo!

You may have heard this term before. "Silo busting" was a hot buzz phrase in business a few years ago. "People talked about 'operating in silos,' meaning they weren't acting cohesively," writes executive coach and blogger Linda Henman. "The metaphor was apt since the new meaning also meant to communicate you'd have to crawl over a barrier to get a message to someone in another silo."

Silo busting was targeted at people in an organization who get so locked into their own department, specialty or individual "silo" that they were unable or unwilling to see or think about connections with other silos. Busting these boundaries, the thinking went, would help stimulate interchange and interaction with different parts of the organization, leading to new, happy and productive collaborations.

Implicit in the act of silo busting is the skill of set shifting. That is what happens in those spaces between the silos: perspectives must be changed and new situations, new people and new and unexpected ideas reacted to. And it's not just limited to the business world. In many areas of science, the greatest discoveries do not emerge in the narrow silo of one specialty but in the cracks between or among specialties—in the spaces outside the narrow field of vision of the specialist.

A way to practice set shifting—the cognitively "light on your feet" skill we've been talking about—is to deliberately try and think like a silo buster. These don't have to be radical, shake-the-world changes, although you may find yourself leaping to new insights that improve the quality of your professional life!

Shift from mental to physical

You've heard me say this before in previous chapters: it's amazing how our bodies and minds support each other. Sometimes the shift we need is to get out of our heads and into our bodies—standing up during a long meeting and stretching, taking a walk around the corridors or the block, doing some yoga stretches or taking a few deep breaths. Many studies have shown the cognitive benefits of physical activity; certainly the "clearing your head" effect is one of them—in this case, clearing and refreshing the mind as surely as it does the muscles. Just as surely as your back and neck feel more supple after that stretch and your legs less tight after a brief walk, so will your mind feel fresher and more flexible—ready and able to shift sets.

At the University of Texas MD Anderson Cancer Center in Houston, employee wellness manager Bill Baun has installed eighteen "stress-busting stations" throughout the Center's facilities for use by everyone from secretaries to surgeons. The stations consist of an elliptical machine, a Precor stretching machine and a special chair with resistance tubing. "We encourage people to take a 'microbreak,'" says Baun, an exercise physiologist and wellness coach. "You don't have to get sweaty; you don't have to do a twenty-minute workout. What we've found is that those who use our stress-busting stations even for just three to five minutes not only release their stress, but also find that that they're more creative, more effective and have more energy on the job."

In other words, just a few minutes of stretching, cardio or resistance training puts them in a place where they are ready to react agilely to new situations and take fresh perspectives on the task at hand or the next challenge coming around the bend. Is there proof of the value to these innovative stations in helping to foster this kind of set shifting and creative thinking? At one point, Baun notes, he and his staff relocated the stress-busting station located on the bridge between the surgeons' offices and the operating rooms. "I had surgeons calling to ask me what happened to their station," Baun says. "'I won't operate before I get on that elliptical machine,' one of them told me. *That's* significant."

Welcome and appreciate the opportunity to shift

Rather than being annoyed or irritated at a call for a shift in your attention, treat it as a welcome messenger or a possibility for new insight and clarity. It's an invitation to rise above the weight of the task at hand—even better, rise above a disorganized life. Greet it with a smile and light energy just as you might notice a much-loved child, mate or pet. Mindfully ask "How can this shift help me perform better?" The answer may not come until later. The shift may be an opportunity not to be missed.

This attitude should be fostered in your personal as well as professional life: Let's say that you regularly spend the holidays at your spouse's parent's house. It's a comfortable routine you've gotten used to over the years. Suddenly, your spouse announces that this year you're going to visit a sister in Ohio instead.

Now you don't mind the sister, but her husband is a bore; they live in a rural part of the state, far from the attractions and stimulations of a city or major town. It's going to be a longer trip—and the weather probably colder. How do you react?

The plodding, lead-footed brain will go along with this change in holiday venues only under duress, complaining and either slow or reluctant to try and even analyze, much less welcome this new opportunity. The flexible, light-footed mind—the master set shifter—does exactly the opposite. It doesn't mean that you have to pretend to suddenly find the brother-in-law's company stimulating or that you must relish a long drive to an off-the-beaten-path location. It's not even a "grin and bear it" attitude. Instead, the set shifter pivots and changes directions, looking at what might be good about this new "set" or preferable to the old one. (Hey, your sister-in-law's a good cook! And did you know they live near a big state park with beautiful walking trails . . . and if it snows, you can rent snowshoes, something you've always wanted to try?)

Make a beautiful decision

Often when a potential shift presents itself, there's a decision to be made. Stop and consider your options: Should you stay at the task at hand or shift to the new task? What are the benefits of either option? Which one wins? Be fully present and awake to the choice. Engage your thoughts and feelings. What do you think about the choice? How do you feel about the choice? Come to a decision; this can all happen rapidly and beautifully.

Don't confuse multitasking with shifting gears

As Dr. Hammerness explained earlier, multitasking isn't about nimble jumping or leaping to engage in new tasks with a shifted attention and focus. It's about attending to multiple tasks simultaneously and mindlessly without deliberate shifting to or from one task to another and back again—or onto something else.

So whatever you do in shifting the set, don't try to do both. Don't try to shift to the new opportunity or task while continuing to attend to the one you were originally focused on. It won't work—both will suffer. Again, think light feet: attend to the new situation and, if necessary, *then* return to the other, hopefully with a new and fresh perspective.

Be a confident jumper!

Whatever the new task or situation—a change in work assignments, a new vacation venue or an interruption that compels you to attend to something else—jump into the new task with both feet, holding the intention of greater performance on the task left behind. Focus with mindfulness and appreciation of the opportunity. Let go of fretting and frenzy. Don't allow yourself to doubt your switch. Trust that the switch will bring new clarity and insight to the newly embraced task as well as the one set aside for a while.

When you return to the task you left behind, stop and pay attention to your mind-set before you shifted and notice where you are now. What is new? What happened to your energy level? Do you find yourself renewed and revitalized? What discoveries have emerged almost effortlessly? What gifts did the shifting bring?

The way to anchor the development of a new skill or behavior, such as set shifting, is to notice the rewards early and often to reinforce what might feel difficult at first. The joy of a new idea, insight or perspective is something we all need more of. It's a key step on your way to the organized life.

CHAPTER 8

Rules of Order/*Connect the Dots*

N OW WE'RE GETTING TO THE IMPORTANT PART. It's time to put it all together, and you are ready. Frenzy has been tamed. Attention is focused. Impulses have been controlled. Memory is tuned. Shifting happens.

You have approached the challenge of getting organized in steps, as individual efforts, following individual Rules of Order. You have taken one positive and organized step at a time and, hopefully, you are already seeing some benefits in your day-to-day life. Practice makes perfect. It takes time to develop a new pattern, a new way of living. Now, we must integrate and orchestrate the discrete steps to yield an organized life—one that goes far beyond your ability to simply remember where you left your car keys.

Before we get into the how-to—and the exciting possibilities of what this can lead to—let's remember our scientific journey and talk a bit more about the organized brain as a whole. We started this endeavor by looking at specific inner regions of the primitive, emotional brain—and the necessity in taming them and controlling powerful emotions that can flood over us and paralyze our thinking brain.

Having regained mastery of our emotions, we then traversed the highly evolved, complex, interconnected areas of the 21st-century thinking brain. You can now visualize regions of your wondrous brain firing away as you hone and apply individual skills, like paying attention and shifting set. Now we want to focus on the incredible circuitry working in concert to make it all come together.

We have referred to the importance of connectivity in brain function all along and, by extension, its importance in the Rules of Order, the skills you can learn and the behaviors you can adopt to better organize your life. But before Coach Meg shows you how to do that, we think it's important in this "pulling-it-all-together" chapter to understand a little bit of how science is also, in a sense, pulling it all together to better understand the overall orchestration of the brain.

In recent years there has been a real explosion of studies on brain connectivity and how the billions—yes, *billions*—of brain cells come together as a working, organized unit via trillions—yes, *trillions*—of brain cell connections. With sophisticated studies and new technologies, scientists are identifying networks in the human brain and understanding how it is organized for large-scale, globally integrated processing. These brain networks allow for maximum efficiency with minimal energy or "wiring cost" (it often involves working smarter, not necessarily harder). Scientists also describe highly connected hubs in the brain, especially critical for the whole network to function. In its organization, the brain has the remarkable ability to perform small-scale, local processing as well.

Several proposed brain networks relate to our efforts to be organized.

The first is an alerting network, responsible for keeping us awake and vigilant to new information and opportunities. This is very clearly a critical basic network, the one that "activates" us, helping us to be prepared and poised for action.

Second, there's an orientation network that takes this "alerted you" and enables you to mobilize your resources to respond—or "orient"—yourself to the new information and stimuli.

And then finally, the executive control network gets involved; these are the brain areas in charge of your thinking, feeling, action—your responses. This network includes brain areas like the anterior cingulate cortex and frontal cortex—regions we have talked about before as being important in efforts to pay attention, focus and so forth.

The three networks function almost as a ready-set-go mechanism. Like a sprinter in a race, the alerting ("ready") network gets you poised and prepared for the race that is imminent; the orienting ("set") network puts you down into your sprinter's "crouch," preparing for action and orienting to the next critical sound: the firing of the starter's gun. The executive control ("go!") network springs into action at the sound of the shot, pulling it all together—the mechanics of your stride, the proper form, the awareness of who's in the next lane and what your competitor's strengths are—as you race down the track.

What's also important to keep in mind as we seek to understand the way these networks function is their interconnectedness. So we can also imagine them as systems strung together, not unlike the way various parts of a major U.S. city are interconnected through cables and wires—all with crisscrossing signals back and forth, keeping us in constant communication and responsive.

Of course, these analogies attempt to simplify a highly complex process—and one that we don't fully understand. Still, while there is much mystery remaining about these networks and how they interact, some answers may soon be forthcoming: in the fall of 2010, the National Institutes of Health awarded grants totaling $40 million to map the human brain's connections using the most powerful brain-imaging techniques available. The Human Connectome Project will yield insight into how brain connections underlie brain function.

These initial grants will support two collaborating research projects led by researchers at Washington University, St. Louis; the University of Minnesota, Twin Cities; Massachusetts General Hospital/Harvard University, Boston; and the University of California, Los Angeles (UCLA). "We're planning a concerted attack on one of the great scientific challenges of the 21st-century," explained Washington University's David Van Essen, PhD, who coleads one of the groups with Minnesota's Kamil Ugurbil, PhD. "The Human Connectome Project will have a transformative impact, paving the way toward a detailed understanding of how our brain circuitry changes as we age and how it differs in psychiatric and neurologic illness."

Said Michael Huerta, PhD, of the National Institute of Mental Health, who directs the National Institutes of Health connectome initiative: "On a scale never before attempted, this highly coordinated effort will use state-of-the-art imaging instruments, analysis tools and informatics technologies—and all of the resulting data will be freely shared with the research community. Individual variability in brain connections underlies the diversity of our thinking, perception and motor skills, so understanding these networks promises advances in brain health."

The Washington/Minnesota team plans to map the brain connections (aka connectomes) in 1,200 healthy adults. This project will examine the influence of genes and the environment on brain connections, as the sample includes identical and fraternal twins. Researchers will look at brain activity at rest and when subjects are performing tasks—such as the ones we have discussed in prior chapters—using a connectome scanner, which will incorporate new imaging approaches, ten times faster imaging and enhanced clarity of images. Moreover, the brain scans will be complemented by movies of millisecond brain electrical activity using a technology called magnetoencephalography (MEG). That's a mouthful, I realize, but it's a remarkable new technology: MEG is a noninvasive procedure, in which subjects lie on a bed while wearing a helmet-shaped device containing magnetic field sensors distributed in a grid over the inner surface. These super-sensitive sensors are able to detect the magnetic fields produced by brain activity and may give a more complete and higher resolution image of the brain in action.

The Massachusetts General Hospital/UCLA consortium will be using a brain scanner system four to eight times as powerful as conventional systems. The team—Bruce Rosen, MD, PhD, and Van J. Wedeen, MD, both of Massachusetts General Hospital and Harvard, and Arthur Toga, PhD, of UCLA— will use brain-scanning technology to map the brain's fibrous long-distance, white-matter connections. In order to best understand the brain's connections, each fiber connecting brain cells and their paths must be mapped. The Massachusetts General team has pioneered a brain-scanning technique that identifies these paths, according to the movement of water along them. This technique, called diffusion spectrum imaging, enables one to make an image of the myriad brain connections and also shows where fibers overlap with each other in an amazingly complex web of cognition.

At the same time as these two projects in the United States, others in the international scientific community are beginning to demonstrate unprecedented, worldwide scientific collegiality in order to help unlock some of the mysteries of brain connectivity. The launch of the 1000 Functional Connectomes Project (FCP) in December 2009 by leading brain imagers has been heralded by many in the scientific community as marking a new era of discovery for human brain function.

This project will allow for unrestricted data sharing among scientists. It began with the inclusion of one thousand brain-imaging data sets collected from dozens of centers around the world. With it, scientists from all over the world can literally and figuratively put their heads together to create the definitive map of the functional networking and connections of the human brain—aka the "connectome."

It appears as if we are entering a time when we can anticipate enormous advances in our understanding of the most complex and remarkable of human organisms. With your reading of this book, you are poised to understand many of the discoveries that are about to emerge. But at this moment, the most important connections you need to make in this final Rule of Order are behavioral. You're ready to pull together and harmonize the skills that we've talked about in previous chapters and that you've practiced. Coach Meg will show you how to do this using some of *her* case studies in organizational success.

CONNECT THE DOTS: COACH MEG'S CASE STUDIES AND SOLUTIONS

What is the ultimate goal of the organized brain?

Is it to lead a hyperefficient life where one is always on time and never late; a life in which every moment is productive and accounted for, where no one ever wastes time; a life spent with a home that is pin-neat, with a garage in which everything is labeled and stored in its proper shelf and receptacle?

Of course not.

Even if those were realistic goals, is that really how we want to live? (Well, okay, maybe the organized garage.)

The real goal of the organized brain is to be able to see the big picture and act on it—living from a higher plane of order. Think about it: in the important domains of our life—such as work, home, relationship, friends and community, our personal health—we're mostly stuck in the weeds when we're disorganized—gets tangled in the underbrush of missed opportunities, poor planning, ineffective communication and ceaseless distraction.

When we are organized and on top of things, the chaos of day-to-day life—and let's face it, there's always going to be some kind of disruption or chaos, in at least one of those domains—is greeted, handled, fielded. We might not be able to change the fact that the car broke down or that the department's budget got cut or that our daughter missed an important soccer practice. But when we're organized we can rise above it; we can better deal with these crises, no matter how small or large. We can better roll with those punches and do what it takes to ensure that some—not all, but some—of those miscues and moments of chaos will not reoccur. And we can do it without having tantrums and without making the situation worse and pushing ourselves deeper into the weeds, deeper into a spiral of further disorder.

Each domain has different degrees of organization. Many people are better organized at work than at home because in the office they are expected to perform, they're accountable and they darn well *better* be organized. But while you may have a neat and tidy desk at work, you may also have an expanding waistline and notice various aches and pains . . . but, well, you never seem to find time to exercise or see a doctor. Or you could feel a spiritual void in your life . . . but, gee, you never can get it together enough to get to church on Sundays or find a few moments here and there to pray or meditate. You are often testy with your spouse, but you don't really seem to get a chance to enjoy each other's company anymore . . . because, hey, we're too busy working and raising kids!

Those are all the fig leaves of disorganization. No one can have it all and do it all. No one advocates a life of robotlike efficiency. That's not what this book has been about. But we can make sure that we see the big picture, in every domain of our lives.

One of the key points we've made throughout this book is that your brain is wired for organization and that you can, in essence, learn from yourself, learn some of the skills and tap into some of the abilities that you already possess in order to become more organized, more in control and less overwhelmed in every facet of life.

If you've been following along, and working on assimilating some of these skills—through our tips and suggestions in the previous chapters—you should be getting to a point where you are ready to put it all together—ready to take the leap out of the weeds and up to a much clearer vantage point.

How we put those first five Rules of Order together—synthesizing them in order to propel us out of the weeds of confusion and into the calm, clear, blue skies above the chaos—is the focus of this last part, this last Rule of Order. And we depart from the structure of the earlier part of the book by offering you my case studies.

Yes, as a coach, I work with individuals too, many of whom have the same concerns about disorganization in their lives. The difference is that Dr. Hammerness's patients are coming to him with issues that often have their roots in early childhood, whereas many of my clients are responding to current challenges in one domain of their lives. They are often at impasses where things seem to be spinning out of control.

I help them the way we have worked to help you in this book—by explaining to them the Rules of Order and then putting them all together, by giving them wings and confidence and by providing them with the tools to achieve the ultimate goal of the organized mind, which is the organized life. That is a life in which you have a sense of purpose and a sense of what you're trying to achieve, a life in which you see the moving parts and can keep them working together harmoniously, a life where you make good decisions and sound choices on most days. A life in which you soar—and flourish.

CASE STUDIES IN THE ORGANIZED LIFE

Megan: Banking on change

Megan is thirty-eight and a rising star at a Boston-area public relations firm. She has two children, ages five and eight, and a husband who works in the insurance business. Right away, there are the pressures of a household in which both parents work. Moreover, Megan recently committed to a charity run, agreeing to raise money for breast cancer, which her mom has been battling. She has to raise money and train to do a five-mile run—which at the moment she's not in shape to do.

When I first met her, Megan had just received a big job promotion, earning the title of vice president. Normally that would be good news for anyone, but she admits that she's beginning to wonder if she should have accepted it.

"Maybe I should have said thanks, but no thanks," she said ruefully.

"Why?" I asked.

"Because I don't think I can handle it. My daughter is in kindergarten; my son's in third grade. They're both really involved with after-school activities, like every kid these days. So I've got a lot to do with them. Plus, I've got my mom. And the fund-raising and the run. And of course, Bill needs me. Meanwhile, now I'm supposed to be attending more meetings, supervising more people and running bigger projects?

I don't feel like I can keep it all together. Some mornings I just don't know what to do first. Do I boot up the computer, open up the box of Cheerios or put on the running shoes?"

As she talked, Megan grew more and more agitated. She was obviously being pulled in many directions and feeling overwhelmed. She also told me she hadn't been sleeping well—or nearly enough, as she worries about what she neglected to do yesterday and what she's going to try and get accomplished the next day.

Megan needs to get organized!

Phase One: Master the First Three Rules of Order

I explained to Megan that the first thing she needed to do was find a way to tame the frenzy.

After a lot of detailed questioning about her lifestyle and habits, we came up with some strategies to help her sleep. This will help her in many ways, not the least of which is that she'll more likely be able to remain calm if she's well rested.

No caffeine past a certain hour. (Megan sometimes liked to have coffee after dinner, and this was not helping her with sleep. We suggested that she make it decaf instead!)

Once the kids were tucked in, instead of going on the computer or having her customary two glasses of wine, I suggested that she turn off her computer and cell phone at 8:30 pm, go into her room, and do ten minutes of deep breathing focusing on her heart and not her head.

Then she could put her "legs up the wall." Anyone who has taken yoga classes knows this move. You simply lie on your back with your legs perpendicular on the wall. Do it for five minutes right before you go to bed, and it will help you get to sleep.

Megan also tinkered with her libations and refreshments. No more chocolates at night, no more coffee, and she cut down her wine from two glasses to one. Too much wine can lead to awakening in the middle of the night.

Now she is more relaxed and sleepier when she gets to bed. And we suggest that she get in bed an hour earlier. Megan tells me that she's been getting an average of five to six hours of sleep per night. The optimum level needed by most people is seven to eight hours.

It pays to pay attention

Our first goal is to tame the frenzy, and part of the way to do that is by helping Megan enjoy quieter nights and more and better sleep. Now, we want Megan to make a conscious effort to sustain attention more fully, mindfully and for longer periods of time. I suggest that Megan try doing something that has worked well for me.

When she gets to the office, now hopefully a bit better rested and less frenzied, don't flit about like most of us do, gnatlike, from one thing to another.

We coach Megan to schedule no meetings before 10:00 am, if possible, and to focus on what she perceives as the most important task of the day (fully aware that there are some days this can't happen, when the client or her boss will demand an 8:30 am meeting or some crisis or new development will emerge). But on the days that she can control her schedule and to-do list, we want Megan to come into the office with a deliberate plan.

No meetings, no checking e-mail first thing, no being distracted by office chitchat. There's time for that later. First thing, I tell her, is to put down the other stuff and focus . . . really focus on the task you deem most important.

For Megan, it turned out to be learning about her new client's business and its past marketing and public relations effort. This is what people call a "situation analysis," and as the name suggests, this means to examine in detail and depth the organization's problems and opportunities, in terms of the way it is communicating (or not) with its key publics.

Analysis, by its very nature, demands focus, so it's the perfect opportunity for Megan to begin mobilizing attention, which I suggest that she do—again, blocking out the first sixty to ninety minutes of the morning, ignoring e-mails, letting the phone ring, keeping the gossipmongers away. I want her to just *focus* on that situation analysis—which involves reading a great deal of material and (that undervalued commodity) thinking!

Her focus time is the morning; for others it may be the end of the day or right after lunch. Exactly when the period of "focused focusing" occurs is not important. What is important is that you learn to put everything else down, turn your attention on something and keep it there.

You don't have to be rude about this either. A closed door and a few words to your administrative assistant, office mate or spouse ("I hope you don't mind, but I need to concentrate on this for a little while...") should suffice. Remember—everyone else is dealing with the same kinds of distractions, so they know what a precious commodity focus time can be.

Hit the brakes

Now Megan has adjusted her prebedtime schedule and modified her habits just slightly. As the frenzy begins to quiet, as she wakes up more fully rested, she is more easily slipping into her morning routine of "full attention" on her new client, the rewards of which are already being reaped, as she's learning a lot about the bank and feeling more confident that she and her colleagues at the agency will be able to solve some of the communication problems facing the client.

Her third piece here is to resist impulsive behavior—to exert that inhibitory control that is at the core of our Rules of Order—apply the brakes.

This goes hand in hand with her focus time, but as she's making a conscious effort to sharpen this skill, she sees it spilling into other aspects of her life. She practiced a mindful response to interruptions, taking a breath and then saying to herself "This interruption is calling for my attention. I get to choose my response. I'm putting my own hand on my shoulder and providing wise advice. Right now the best choice is to set the first focus aside until later in the day." Occasionally the interruption is urgent, and Megan pauses and chooses to step into another focus and returns later in the morning to resume the situation analysis by rescheduling her day. The act of creating a moment of pause to allow her internal counselor to speak makes all the difference.

She no longer feels compelled to immediately attend or respond to every e-mail. While working on her situation analysis in the morning, she will occasionally see e-mail previews pop up on her screen. In the past, that's all it took. Immediately she was off topic and on to a response to the e-mail. Now she keeps her attention on the main task.

It's the same thing with the phone, with the office gossip girl or even the friends who come by and want to ask her if she saw who got voted off *Dancing with the Stars* last night.

Megan knows how to do this politely—she's not slamming doors in people's faces—but her external demeanor is as important as her internal reaction to potential distractions. Her mind is now getting better at putting the hand on her shoulder and gently returning her to the client's work.

There will be plenty of time to respond to e-mails or chitchat with colleagues later.

Oh, and a funny thing has happened, Megan reports to me during our biweekly phone coaching session that while she's been practicing her attention and inhibitory control skills at work, she's noticed that things have changed at home, too. Megan can now focus on making dinner, having a conversation with her mother on the phone or helping her daughter with homework. A noise in the other room, a phone ringing or a dog barking does not pull her off task the way it used to. She is learning to acknowledge the other stimuli and evaluate their importance. ("Was there a crash or the sound of something shattering? No? Okay, good. Whatever fell can be picked up later!") Instead of running pell-mell into a new crisis (whether real or imagined), she now sticks with what she was doing, therefore accomplishing more and feeling less frenzied, which in turn (and along with her lack of caffeine, reduced wine consumption and deep breathing and yoga moves) is allowing her to sleep better.

Phase Two: Now We're Cooking!
Set Shifting, Working Memory and the Organized Life

When Megan first came to see me, she didn't really seem to know how to handle all of life's demands simultaneously. She was constantly worrying about the neglected domain when attending to another domain—her job and the promotion, her kids, her marriage, managing the household, supporting her mom and allowing time for her outside activities such as the fund-raising and training for the run.

She clearly didn't want to fail in this new responsibility she'd been given at the office, but on the other hand, she didn't want to neglect time with her kids. There was a huge inner conflict.

One thing we did was to help organize her priorities by thinking strategically—and applying the two "higher order" Rules of Order which, as you'll recall, are the ability to mold information (working memory) and cognitive flexibility (set shifting).

I'm not a therapist, I'm a coach—a specialist in change—but in order to change her behavior we did have to talk a little about what was underlying her anxieties. She realized that the key to managing her overwhelming feeling was handling her office job (as opposed to her other "jobs": mom, spouse, homemaker, etc.). The other aspects of her life were clearly important, and they were adding to her sense of distraction and disorder, but concerns about her job performance seemed to overshadow the others.

She decided to make her organizational challenge work-focused—specifically the new client she had at work, a Boston area bank.

The feeling was if she could get on top of this project, she'd feel better about her job. And that in turn would help her feel better about (and more in control of) other domains of her life. In other words, if she wasn't coming home worrying about what she hadn't gotten done on the bank project at work—if instead she came home a little more calm and feeling as if she'd made progress—she'd be more available to the kids and to her husband and would go to bed more rested.

So the decision was made: and now, with some of the first steps successfully taken—frenzy tamed, attention sustained, distractions on the job managed—she's ready to get to the most exciting and rewarding part of personal organization.

The development of the "higher level" skills of working memory and set shifting will not only help you get better organized, but they will also make you more effective and improve the quality of whatever you're doing.

In Megan's case, figuring out how to give the local bank a better image.

The bank had problems, as almost every bank in America has during rough economic times. Surveys showed that everyone hated the bank; they hated the checking and ATM fees and the fact that they seemed to have given out loans too easily before the housing crisis—and too reluctantly afterwards. Some older customers still resented the fact that they no longer used "passbooks" for their savings account. Younger customers complained that there weren't enough ATMs or that the website wasn't fast enough for them when they wanted to do their online banking.

The bank executives had presented these issues in a meeting with Megan and her bosses—who promised the client that they would fix everything; "Just leave it to Megan," they said. ("Great," thought Megan. "They left it to me!") At first—and this was about the time she decided to call me—she didn't know what to do. The problem seemed intractable. Financial institutions across America were in a crisis of historic proportions. This bank was probably lucky not to have shuttered its doors. The economy was coming back but not nearly fast enough. How could she possibly change all the negative attitudes toward her banking client in this kind of climate—and with all these factors that were well beyond the control of a public relations agency?

I certainly didn't know the answer because I'm not a public relations practitioner. However, as we just discussed, in our efforts to get Megan better organized, we did suggest the morning focus hour, where she cut out a block of time every morning in which distractions were avoided as best as possible and full attention paid to the problem at hand.

There, as she read and thought and began to analyze and fully understand the client's situation and outside trends in the industry and among consumers, a cooler and calmer Megan began to realize that she could indeed help this bank.

Assembling ideas, achieving insights

Megan now needed to build a vision for the campaign she was going to propose to the client. She needed to assemble the working memory "pieces" that would provide both the foundation and the spark of the big idea for the campaign. She jotted down ideas from her meetings with the client, ideas from some suggestions her boss had made, and, of course, she had a bunch of thoughts that had popped into her head during her quiet focus time—and even a few things that she'd thought of while on the treadmill, preparing for her five-mile run. She also held a brainstorming meeting with her staff to generate and collect more ideas that way.

My contribution to this was to suggest to Megan that she collect all these ideas and physically lay them out in front of her. By studying these options, ideas, fragments of thoughts—whether on a piece of paper, in a scrapbook or in a file in her computer—Megan is *building a rich working memory.* The information here can now be molded and examined. It's from these pieces that the big idea can come.

(As a quick aside, this technique is used in other fields as well. Roy Peter Clark, a nationally known writing coach from the Poynter Institute—a journalism think tank in St. Petersburg, Florida—recommends what he calls "composting" as an important step in the writing process. By this he means gathering any pieces or scraps of information that might be related to the topic you're going to be writing about and keeping them together in one place, physically or digitally. What he is advocating is exactly what we're talking about here; a marshalling of information that can later be examined, considered, "molded." It's a sort of woodshed filled with kindling that will in turn provide the spark for the idea that will "light up" the story or the article.)

So now we want to foster insight. Now we want to get to the place where an original idea arrives. This will require being ready to jump on the opportunity, as the insight is sparked by the molding of your information. This is, as we discussed in an earlier chapter, set shifting.

I wasn't looking over Megan's shoulder to see the lightbulb go on over her head. It might have happened while she surveyed the notes and pieces on her desk. It might have happened when she was out for a three-mile power walk, training for that fund-raiser. It could have even happened when she was puttering around the house or relaxing with her legs up the wall.

The point is it happened. Memory provided the spark, which enabled her to demonstrate some cognitive flexibility—to set shift—into a whole new way to look at the problem.

The result was not one, but *two* big ideas.

The first was an "image" ad campaign, in which real customers of the bank—individuals who had been able to put their kids through college, thanks to some suggestions made by the bank's financial manager or who had saved their homes because the bank was willing to restructure their mortgage—talked about why they were loyal. Apparently, this kind of thing had never happened before—but during one of her morning focus sessions, Megan came across a file, almost lost among all the background material the bank had sent her, that contained a number of e-mails from customers. A surprising number of them were positive, even heart-warming. That became one of the "pieces" of her working memory; when she remembered an ad campaign for a credit card company that talked about celebrities who had used the card for years, the set shifted. The bank, like most, advertised in local media. But maybe instead of just advertising rates, why not advertise relationships with customers?

She had a second idea as well. The client had told Megan they were interested in trying to get more college students—of which there are many in the Boston area—to open up accounts with the bank. Megan assigned one of her junior staffers to research the ways that banks around the country were trying to communicate to this audience. The staffer sent her links to a couple of Facebook pages from other banks.

This, too, became a piece of working memory.

The idea Megan had was to create a Facebook page, sponsored by the bank and with their special student rates advertised in a banner ad on the side, but one in which the emphasis was not just on talking about how great the bank was but rather the sharing of financial tips and advice among young people. An expert on this topic that Megan had found by Googling the topic (another piece of the working memory) would periodically post some tips on financial management for young people, and the bank could provide free, no-hard-sell seminars for students.

The insight for all this came from seeing connections between things that were previously unconnected: Advertising as a vehicle to tell customers' stories, instead of simply selling bank products. Social media to reach the student population. A Facebook page to provide information as opposed to just trying to "sell" them. This is the building of new connections in the brain. This is thinking out of the box. This was not just being more efficient, but being more innovative and more creative.

That's what happens when you really get organized.

Let's be honest: if Megan wasn't talented and experienced in her career and if she hadn't had a good staff, a more-or-less supportive boss and a client willing to listen to new approaches, none of this would have happened. But along with those prerequisites came Megan's newfound organizational skills. This strategy and the subsequent campaign (which went on to win some awards from the Boston public relations professional association and eventually garnered Meg a nice, fat bonus) might not have come off so well, or even at all, if frenzy wasn't tamed and if

Megan didn't have a sense of focus and traction in her day. The ideas would probably not have been nearly as dynamic had they not been the product of a rich working memory or a nimble mind ready to shift and make new connections. And she pulled it all together. That's the feeling of an organized brain, not necessarily one that knows exactly where everything is in your closet, or that can account for every pen in your desk drawer—although all those things can help. What we're really talking about here is the brain that knows the steps to take to achieve the goals that are important to you; a brain that uses the Rules of Order—first to sweep clean and calm the chaos and then to achieve productive, new problem-solving insights.

That's the organized brain. I was so proud of Megan that she used hers to achieve a great professional success—one that, in turn, enriched the rest of her life as well.

Stu: Taking time to improve home and health

Stu, a forty-five-year-old sales executive and father, needed me to help him make some positive changes in his lifestyle.

He had heard that I was an executive wellness coach and, I think, had at first assumed that I was some sort of personal trainer who would take him to the gym and show him how to do a proper push-up. I explained to him that I was really there to assist him in making the cognitive and behavioral changes he needed and wanted to make.

What I was also doing was helping him get his brain organized.

Because, for many, it is not just laziness or lack of motivation that keeps them from following a regular exercise program or healthy diet. It's disorganization. The most common reason cited for not doing these things that we all know need to be done is lack of time. "I don't have the time to get to the gym. I can't possibly find the time to prepare healthy meals."

I hear this complaint about lack of time all the time.

It's really the cry of the distracted and disorganized, and I could immediately see that Stu was one of them.

There were a number of things he talked about at our first session, in addition to his health. He told me about some home projects that he started and never seemed to finish and how his wife, Diane, was getting angry at him about that. It was also Diane—not to mention Stu's primary physician—who urged him to lose some weight and start eating better. When I met him, the football-sized bulge in his shirt suggested that Stu was carrying some extra pounds. He sat down gingerly in the chair in my office—complaining about a sore back and a bad knee and was absentmindedly sipping soda from a cup emblazoned with the logo of a well-known fast-food franchise.

When I couldn't help looking at that logo, Stu hung his head theatrically, like a kid caught with his hand in the proverbial cookie jar. "I know, I shouldn't be eating fast food, I'm a bad boy," he said. I had to laugh. Stu was charming and funny. I could see that he had probably used these attributes to avoid the things he knew he was supposed to be doing.

Stu had a lot going on in his life, like most of my clients. Business was challenging; there were big clients to please and new technologies introduced to his office that he felt pressured to master. But I got the sense from talking to him that the thing that was really causing him stress right now was his relationship with Diane. Again, I'm not a counselor and I certainly didn't try to offer marital advice, but it was clear that Stu felt that his wife was constantly, in his words, "on my back" about finishing the home projects and about getting himself into better shape.

I got a feeling that if he could get a handle on those things, he would feel better and more energetic, his marriage would improve and then he could function better on every front.

Tame the frenzy at home, so the big picture at work becomes clearer

Frenzy presents itself in different ways. Megan's was almost palpable. Stu's frenzy—like a lot of men—was quiet, balled-up tension that occasionally was released in a spate of yelling, and occasional kicking or throwing of things or drinking a few beers. Despite his congenial exterior, I also could tell that inside Stu was worried about what the doctor told him. His father died of a heart attack in his early sixties, and he was afraid that he was traveling down the same road. He was dejected because of the extra pounds he was carrying around. He lacked energy, didn't like the way he looked and couldn't get things done as effectively as he used to, which made him feel even more under the gun and disorganized.

Our overall strategy for Stu, then, was to help him get a better handle around two of the most important domains in his life: his home and his health. This will help tame his frenzy in several ways. Diane would be pleased, and a healthier diet and physical activity regimen will simply make him feel better.

Stu needed to find time, and—since his wife is one of those pushing for him to make changes—I suggested that he and Diane do some time bartering, working together to come up with a better timeline, starting with Saturdays.

He needed to carve out time for exercise—so Stu suggested that he get an hour first thing Saturday morning to exercise. He'll get up and go for a walk in the nicer weather or go up to the local gym in the winter. After his workout, instead of stopping at the nearest fast-food joint for some junk food, he agreed to help Diane prepare the healthy breakfast that they both need—and that they will eat together (and—bonus!—maybe even persuade their teenage daughter to try; although at 9:00 am on Saturday their fifteen-year-old is usually still sleeping). Stu agreed that he will devote his energies after breakfast to a home project.

While it's not a cognitive skill like the rest of our Rules of Order, don't underestimate the effect of nutrition in helping with your organizational issues. With a breakfast of egg whites, oatmeal and a piece of fruit, Stu will be making his doctor happy, but he'll also be fueling his brain. The brain frenzy from a high and fast dose of sugar is terrible. You also feel guilty from eating that donut, you don't feel satiated, your head aches and you crash later.

To keep up his part of the bargain for Saturday morning, Stu realizes that he will have to change his Friday nights. He and the boys at the office had a long-standing tradition of meeting at a local watering hole after work, to drink a few—a "few" as in three, maybe four drinks. Now he's going to have to cut that down to one drink and probably leave the bar early, if he even needs to go at all.

After the first two weeks of the new schedule, Stu admitted to me, "I felt better when I got up last Saturday. I was less sluggish, more clearheaded." That was due to the reduction of alcohol and the fact that, by getting home earlier after work, he was also getting into bed a little earlier.

Meanwhile, they're tweaking the schedule, with me on the sidelines—like a real coach should be, offering some reflection, encouragement, brainstorming and the occasional suggestion. He and Diane decided to spend ten minutes after breakfast each week planning how he's going to use the task hour (during which Diane goes to a Jazzercise class at a studio near the house—so she's getting her exercise in the morning as well). It's a collaboration—which also means they're cooperating instead of fighting over this. Instead of conflict, they're in a partnership. This, I know, is reducing Stu's tension and frenzy.

What about our second Rule of Order, sustaining attention? We are achieving this through the scheduling and allotment of certain times for certain things. Each hour has its purpose. Now Stu has focus; distractions are controlled. When he (finally) fixes the sink in the guest bathroom or paints the garage, he will do so without being distracted by worries over upsetting his wife or tension over his sedentary, junk-food eating ways. He can apply the brakes to any impulses to plop on the sofa and watch television, because he knows that—thanks to his more organized, consensually structured morning—he will have time to do that later if he wants.

He has a plan now; he's doing something about the issues that were overwhelming him. He's tamed the frenzy. He and Diane are cooperating, not crossing swords, over things around the house.

After about six weeks of this, Stu dropped by my office. He'd lost weight, he looked better and he felt better. This was obviously because of his exercise and improved diet and healthier lifestyle—but that was all made possible through his and his wife's organizational efforts. And, he told me, they've added to that agenda. They've decided that on Sundays they'll walk together first thing in the morning, after which they'll go to church as a family—something they often missed because Stu was usually running around trying to finish up the home projects he had left unfinished the day before.

After church, he told me, he and his wife have decided to have lunch together at a local restaurant that has an extensive "healthy choices" section. There, Stu said, they plan to review the progress of the home plans, which, I noticed, he seemed to be speaking about with greater enthusiasm than when he first came in to see me (he excitedly talked about the new home office they're creating out of a spare bedroom).

Even their daughter has noticed. While they haven't persuaded her to get up much earlier—she is a teenager, after all—she's started doing her homework while her dad's working around the house and now calls for dad's limousine service—to take her to her friend's house or to the mall—on Saturday afternoon, by which time Stu is feeling a lot better about himself and his life.

He's so happy he'll even give her some extra cash to spend at the mall.

Moving on up: Stu soars

Stu has made great progress. Keep in mind that at this point he's really only begun to master the first three Rules of Order. It would be perfectly fine if he decided to stop right there. But Stu has decided to keep on soaring. He wants to employ the two "higher order" Rules of Order—which, as I've said, is often where real creative and organized thinking begins. He wants to plan a home office. His boss and two colleagues have long been working at home two days a week and have encouraged Stu to do the same because of the reduced stress and time in commuting, as well as luxurious quiet time for an intense focus on creative projects that require strategic thinking. Given the stress at home, the lack of a suitable home working space and Stu's concern about self-discipline and interruptions when not at work, it has never seemed like a good idea.

Now that Stu's frenzy was tamed, he began to think about the possibility more, even gaining a sense of adventure. One Sunday afternoon Stu asked Diane to help him think about all of the considerations that would help him make a decision. Diane suggested that she could stock the fridge with healthy food that he liked for breakfast and lunch; Stu began to imagine that he could use his commuting time for a morning workout. Stu remembered his boss's trick of setting a timer on a key project to provide urgency and a little self-competition—can I complete the project plan update with just the right amount of detail and share with colleagues in under an hour? (Just as Megan did when she began to envision her bank campaign, what Stu, along with Diane's help, is doing here is to exercise working memory. He's molding information.)

Diane and Stu began to sketch out a layout for an office in the spare bedroom and plan where to put its current contents. Something Stu had yearned for is an office with a window, and that's what his home office could deliver. The savings in gas and parking expenses in a year could pay for a new flat-screen HD television for Sunday football.

Yet at the same time, Stu could vividly picture the downsides. He was so easily distracted. How could he not be sidetracked with temptations at home? Watch television? Run an errand? Deal with people at the door? Talk to Diane? Play with their dog? He would feel worse in the end if they invested in creating a home office and he felt more disorganized and less productive than he already is.

While Stu had stocked his working memory with all of the considerations around working from home, he still felt ambivalent, stuck on the fence. Where would the insight or shift in perspective come from?

As I coached Stu, I helped him think about the pros and cons of the home office. Stu felt motivated—he wanted to be fitter, healthier and more productive. He also liked the idea of saving money on gas and spending less time commuting. However, his confidence was lacking. He had worked in an office environment his whole life, and this would represent a big change. What insight was needed to gain confidence and to see things in a fresh light?

One Saturday morning, he bumped into a buddy at the gym. "Stu, I've been watching you from the other side of the weight room," his friend said. "You've been here going nonstop from station to station for 45 minutes. I remember a year ago you told me that you could never stand coming here, that you got bored after a few sets and quit."

Stu realized his friend was right; he was self-motivated and he could sustain an hour workout now. In part, that was because he liked the results he was seeing (his pecs and biceps were reappearing, neither seen in some years). Plus, Stu had lost a couple inches from his waist; he could now walk briskly for 45 minutes without huffing and puffing. Clearly, he had succeeded in getting himself back into shape and better health habits.

Maybe, Stu wondered, as he related the story to me the following week, he'd get some similarly big results if he developed the focus needed to work creatively and productively at home.

Eureka! That was it: Stu's set shifted. He was now looking at the home office from a new perspective. Stu rose to the occasion; the home office was a stretch for him, just like his workouts had been. If he was successful in working at home, he'd feel at least as good as he did about his pecs and biceps (which were appreciated, too, by Diane).

HOW MEGAN AND STU PUT IT ALL TOGETHER

What we've shown through the stories of Megan and Stu is the real prize of organizing our brains. When we've cleared the frenzy and distraction landmines, focused attentively and assembled our working memory, new insights arrive. The big picture comes into view, as if the clouds have lifted and there it is. That's the moment when we feel as if life's chaos has lifted for at least a while. We are clear about our direction and confident in the process it takes to get there.

It's a wonderful feeling.

Obviously, their situations were different than yours. But I suspect that you can see a little of yourself and your life in some part of Megan's or Stu's. And there are some lessons we can all learn from their experiences, as we try to better organize our own lives:

Start with one domain to work on

It could be your job, your family, your relationship or (as was the case with Stu) the need to take charge of your own health, but pick the domain where you're most determined to succeed—and the one in which you have the most confidence of success (hopefully, that confidence has been buoyed by now having a better understanding of the tools and Rules of Order). That doesn't mean you're going to disregard the problems in your life that may seem more intractable. You will get to them in due time. But we in the coaching business have learned that success breeds success.

Have faith

The progress you make in the domain you choose to work on will spill over to other domains. The more skilled you are at taming frenzy and impulses, or assembling a rich working memory, the quicker and more effectively you'll be able to apply those skills to the next domain you choose to work on.

Approach it as a challenge

While they may have had trepidations at the start and some natural setbacks along the way, both Megan and Stu (and hundreds of other clients I've worked with) began to really enjoy and take satisfaction with the process and the progress they made as they began to get their lives together. That's the right attitude to start with: yes, you may feel frenzied or out of sorts or overwhelmed right now, but as we attack this problem and as you begin to slowly but surely take the steps outlined in this book, follow the tips we've suggested and eventually pull it all together, you should be deriving satisfaction from the journey. Getting yourself better organized is good for your overall health (physical, mental or emotional), and it can be satisfying and rewarding—and, yes, fun, if you approach it that way.

Remember the constant of change

Life is full of surprises, some good, others…well, not so much. Even with all of the work we've done to rise above the weeds to the clear blue sky, peace of mind can be quickly rocked by new events. Enjoy and appreciate both the moments along the way of getting organized as well as the great vista when you get there. It may not last long! Fortunately you now have the skill to get *re*organized; you can follow the rules again to get to the inner peace that comes from moving from chaos to order— a point we will consider more closely in our final chapter.

CHAPTER 9

Staying on Top of a Fast-Changing World

A N ORGANIZED MIND ANTICIPATES AND WELCOMES THE FUTURE. That's what this final chapter is about: how to stay organized in the future; a future that many predict will be so technologically advanced with gadgets so powerful that our minds cannot possibly cope.

Yes, the pundits are already wringing their hands over this. "The online world has merely exposed the feebleness of human attention, which is so weak that even the most minor temptations are all but impossible to resist," wrote one critic in *The New York Times*. Another article raised the question "Is Google making us stupid?" and concluded that our own intelligence is "flattening."

If you are worried that your brain is flatlining and that it may not be up to the tasks of dealing with the next Google, Facebook or Twitter—or even more significantly, with the supercomputers, nanotechnology and robotics that, futurists tell us, are going to be the major technological breakthroughs of the next couple of decades—let me reassure you: you're up to the task.

Yes, your brain and these Rules of Order are powerful, powerful tools. They will carry you far, no matter what the advances in technology. We do not have to feel overwhelmed by the sheer magnitude of information being birthed on the planet. Instead, we can thoughtfully take in whatever we choose, provided, of course, that we're calm, focused, flexible . . . all the things we've talked about in this book.

There has been a lot of imagery used in the earlier chapters to describe various brain functions. The one simple image that comes to my mind when I think of the organized mind is a pyramid of building blocks: each rule building upon itself to get you to new heights. As Coach Meg pointed out in the last chapter, it's a good view from up there, once you get out of the weeds of disorder—a good view not only of your present but to the future and to what is coming down the road.

Having now summited the pyramid of organization (or close to it, I'm sure), you are well equipped to handle the future. Not only do you have the view from the top, but you have each building block in place to work again for you.

As you come face to face with the future, you can count on these Rules of Order again and again. Think of the latest blitz of hype about a new "must-have" gadget—one that will make your life easier and more efficient and you faster, stronger and able to leap small buildings with a single bound. Is this really something that will rock your world? Is this really the "killer app," the game-changer, the PC or iPod or cell phone that will truly change the way you work or play?

Maybe yes, maybe no.

The first response is to take a breath, and consider that it—whatever *it* is—was inevitable. You could have almost predicted it. At this writing, for example, Facebook is the most popular social media tool on the planet, claiming 500 million active users. But at a recent conference, we heard a tech consultant predict that five years from now no one will remember it. I tend to doubt that (reruns of the film *The Social Network* at least will ensure its longevity). Still, we understand the point: that Facebook and Twitter and YouTube and MySpace will be joined by other forms of social media—or even entire new genres of ways to communicate digitally—that we, writing in late 2011, can hardly conceive of.

But it's coming! And when "it" arrives, you don't need to panic. "This doesn't unnerve me," you should say. "I've managed the changes in the past; I'll manage them in the future." Indeed, you might want to think not about how this new technology will make your life more difficult but, rather, how you can use this technology to make life simpler, to enhance it, or to make you more effective.

So when the inevitable changes occur, remain calm and embrace perspective. This is where you use your attention, your working memory, your newfound mental flexibility. You keep your attention focused and your thinking on track. Think about how it seems like just yesterday that you were first introduced to texting, e-mail or mobile phones and how it took time for those technologies to become integrated in the workplace or home. How it was a gradual process, in which early versions of the new technology were improved upon, as major kinks were ironed out. You didn't start sending one hundred texts a day the first time you figured out how to do it. As social researchers have shown, the adoption of innovation takes time…even now.

Engaging your memory and your ability to think flexibly, you think to yourself, "If I did x, then probably y will happen, or maybe z. If instead I started with y, then x…." With self-talk, you run through different scenarios in your head. You are thinking beyond one moment in time; you are thinking, "How have I responded in the past, how did that work and how did that not work for me." This thought process keeps you in charge and enables you to ask yourself, "What does this gadget or technology mean for my future?"

A calm, thoughtful perspective allows you to think really clearly about the latest approaches—whether they be new hardware, a new political candidate or a new mind-set. You will out-think the new gimmicks and see through the empty promises. Maybe you should embrace this new technology—take a tutorial at your earliest convenience and practice at home. Maybe you should use it sparingly. And maybe, just maybe, you don't need it at all! Regardless of which direction you go, you approach the decision thoughtfully, without fear or panic. Your attention is finely tuned and not reactive or impulsive. And it needs to be. Because again, as we've said from the beginning of the book, the world comes at you in a fast-paced, high-speed stream of real and virtual information, zooming right at you and blanketing you. But you don't need to live so fast. You can't live so fast that there is no room for real thought or calm perspective.

That's not a shopworn call to stop and smell the roses. We understand that life for many is too hectic to do that. What we're saying is that before you gun the engine right past the garden, or even worse, drive right over it and flatten the roses, use the tools you've learned—the tools of impulse control, calm, focus—to organize yourself and make a smart decision.

Granted it is sometimes difficult to take the approach we describe, as we are pushed to interact with the world as if it is one giant touch screen; we must flit from one image, one idea, one plan, one piece of information to another. So we need to continue to sustain our attention and use our working memory to mold information and to pause and remember what the goal of our current task is. If we don't work to *keep* our thinking on track, we could *lose* our minds online. The Internet can be a powerful, necessary working tool as well as a diversion. Don't confuse the two.

So before you start clicking away, think it through for a moment or two. Have a purpose or objective before you start searching. What am I looking for? What am I attempting to find or learn and for what purpose? If you went to check a statistic for a report at work, and you find yourself scrolling through the batting statistics of your favorite team or some juicy new celebrity gossip . . . well, you've gone off track. You have become a momentary victim of the Distraction Epidemic! And while you might enjoy reading about Derek Jeter or Lady Gaga, you're going to waste time, and later, as you're pressed for time, you're going to feel more disorganized and more under the gun.

Over time, we will understand more about the influence of the Internet on our brain. Studies are beginning to emerge, again using the latest neuroimaging technologies to examine the brain's activity when it goes one on one with the Internet. One recent "techno-fascinating" study found differences in brain activation according to whether people had Internet experience or not. Brain activation (while on the net) was greater in people with Internet experience as compared to those without net experience; multiple brain areas were turned on, involving areas that have to do with decision making and organization. Is this a sign that with use the Internet can make us smarter, not more stupid? Or just make our brains work harder? Time will tell, but this intriguing study is another example of brain science showing us how the brain adapts and can change with experience.

Overall, regardless of the technology, or what the future brings, you must stay true to yourself and to your goals and aspirations: whether it's your plan for the moment, the week, the year or the next ten years. With this mind-set, you can set shift and embrace new worthy opportunities; these are hallmarks of an organized, 21st-century mind.

LAST WORDS FROM COACH MEG: THE "RE-ORGANIZED" MIND

We've talked about that time when you are able to rise out of the weeds of disorganization and into the clear skies, a peak moment when everything seems to fall into place and you are truly organized, thinking clearly and creatively. It's been the goal of this book to get you to that point.

Just don't count on it lasting too long.

As you've learned from Dr. Hammerness, the brain is a dynamic, changing organism. It is also true of the organized mind. It is not a permanent state, in response to our constantly changing world. The arrival of new upsets, new distractions, new life questions and new sources of frenzy, whether of our choosing or not, is a certainty. Perhaps you've got one domain under control—your relationship with your kids—when an upset occurs at work, your sister develops a health issue or you make a stretch commitment to a community project.

These new forces disturb the picture and brief periods of equilibrium. We gently spiral downward—although not too far—back into the weeds and lose sight of the blue sky for a while. Once again we are called to renew the process: refocus, refresh the commitment to handle impulses and move intentionally through insights to the vista where we can make sense of what matters and what doesn't and be on top again.

So let's revisit what you've already learned and use it to get back on track.

Rekindle your motivation

Recall in Chapter 2 that I explained that developing a more organized mind is a process of change and there are some important ingredients to have in place to make change that sticks. The first and foremost is to find your internal motivation: what is driving you to want to have a more organized mind. Now that you've got a more organized mind, revisit what is driving you and make sure it's still burning hot, maybe even hotter as you're enjoying being organized. Perhaps you are energized by the sense of being in charge and in control of your life. Perhaps your mate and kids have remarked how relaxed you seem and how they appreciate your fun company. Maybe a coworker has noticed that you have gotten more confident and seem to be enjoying work more.

Harness your capacity to adapt and change

Back when you started reading Chapter 2, we talked about understanding your challenges (the "cons" for not changing) and developing strategies to overcome them. By now you've experimented with many strategies to sweep your brain clean and move confidently to the big picture. You inventoried your strengths, such as curiosity or love of learning, and you put them to work in service of your dream of an organized life.

Over time you've come up with a nice mix of habits that work to sweep away the frenzy, stay focused, populate your working memory and make leaps of insight. You practiced until they became automatic. Notice your enhanced capacity to experiment, create solutions and practice until new habits and brain pathways have formed.

You now have what it takes to make the best choices moment to moment. You have a lighter touch; you've become more nimble.

Connect with your body

You appreciate now more than ever that it's not all mental; the physical stuff is hugely helpful. If you've let go of these habits during a new period of change and disruption, now is the time to get back into it: sleeping better, exercising three times a week, meditating before you go to bed, eating lean protein at breakfast for brain energy or a bowl of blueberries when you need an antioxidant boost and a quick ten-minute walk when you need to recharge. They will help you in so many ways, not the least of which will be your ability to get yourself *re*organized.

Appreciate how good focus feels

Remember when you struggled to stay focused on a task? Now you enjoy multiple focus episodes during your day and have started to relax into them and enjoy the ride without guilt or fretting, like getting on a bicycle or motorcycle and riding carefree on a sunny day. Enjoying an activity for its own sake. Ahhhh....

Get your heart and head working as a team

Now your emotions aren't strangers; they are friends and full of insights. When you pay attention, their wisdom helps you turn on a dime or stay on task, whichever is the best choice. You welcome your impulses and their gifts rather than resent them or jump on board mindlessly.

Launch all systems

With unfrenzied focus, you now amaze yourself with your new skill in consciously assembling all of the bits of working memory you need for any key task. How good it feels to have everything you need at your fingertips and to look out at all the things to keep top of mind! What fun it is to let go of the task with intention and jump into a new activity knowing that ideas will pop into your head when you least expect them to arrive. The pleasure of new insights arrives in a burst, like a delicious mouthful of your favorite food; your senses are awake and you feel alive.

Enjoy the view

If you've ever hiked up a mountain you know what it feels like to stand at the top and look in many or all directions, in awe of nature's beauty, and to feel at one with the universe. When you arrived at the top of your last mountain, on top of the domain of life that you strived to organize, you appreciated the moments of awe and beauty. Remember those moments. The memories will keep you going as you move up the next mountain of your life.

Humans are wired to deftly handle the ever-shifting winds of life. One storm or challenge passes and we pat ourselves on our backs for a job well done. We're on top for a bit—but only for a moment before the next challenge to our well-earned picture emerges.

A recent study by psychologist Barbara Fredrickson determined that the factor that most distinguishes people who are satisfied with their lives from those who are not is their higher level of resilience. Your successes so far in following the Rules of Order—whether it's led to small or enormous changes in your life—should help give you the confidence you need to bounce back from and adapt to whatever comes along and rise to the big picture sure-footedly and swiftly.

This is a powerful way to create not just an organized life but a life you love.

The Rules of Order At-A-Glance

In *Organize Your Mind, Organize Your Life* we boil down many essential "brain functions" to six principles—what we call the Rules of Order. Consider these "brain skills" or abilities that you can develop and master.

1. Tame the Frenzy: Organized, efficient people are able to acknowledge and manage their emotions. Unlike many who let their emotions get the better of them, these folks have the ability to put the frustrations and anger aside, almost literally, and get down to focused work. The sooner the emotional frenzy welling within you is tamed, the sooner the work is done and the better you feel.

2. Sustain Attention: Sustained focus or attention is a fundamental building block of organization. You need to be able to maintain your focus and successfully ignore the many distractions around you, in order to plan and coordinate behaviors, to be organized and to accomplish something.

3. Apply the Brakes: The organized brain must be able to inhibit or stop an action or a thought, just as surely as a good pair of brakes brings your car to a halt at a stoplight or when someone cuts suddenly into your lane. People who don't do this well will continue to act or think in a certain way despite information to the contrary.

4. Mold Information: Your brain has the remarkable ability to hold information it has focused upon, analyze it, process it and use it to guide a future behavior, even after the information is completely out of visual sight. It is capitalizing on working memory, a kind of mental workspace.

5. Shift Sets: The organized brain is ever ready for the news flash; the timely opportunity or last-minute change in plans. You need to be focused but also able to process and weigh the relative importance of competing stimuli and to be flexible, nimble and ready to move from one task to another, from one thought to another. This cognitive flexibility and adaptability is known as set shifting.

6. Connect the Dots: The organized and efficient individual pulls together the things we've already talked about—the ability to quiet the inner frenzy, to develop consistent and sustained focus, to develop cognitive control, to flexibly adapt to new stimuli and to mold information. The organized and efficient individual synthesizes these qualities—much as the various parts of the brain are brought together to perform tasks or help solve problems—and brings these abilities to bear on the problem or opportunity at hand.

The Top 10 (Dis)organizational Complaints—and Our Solutions

Think disorganization and what's usually the first example people will come up with: car keys left on the kitchen table, sunglasses forgotten on the counter of a convenience store or an umbrella abandoned under the table at a restaurant.

Good old-fashioned "forgetfulness." There's nothing new about that. However, the almost quaint aspect of such behavior—the idea of the "absentminded professor"—belies the fact that in a world with greater choices, more technological options and increasing complexity in almost every aspect—social, work, family—the issue of disorganization and inattention has become serious enough to warrant its own label among the parade of public health crises bedeviling society today: the Distraction Epidemic.

In the preceding chapters, we have talked about the new advances in our understanding of how the brain is organized, in order to give you greater insight into how you can tap into some of the intrinsic tools and skills available to you—our Rules of Order—in order to better organize your life.

However, there may be times when you face very specific organizational challenges. Hopefully, by the time you read this, you will have learned and practiced some of our "big picture" skills; our Rules of Order. But sometimes you might just want a quick-fix, at-a-glance approach to dealing with some of the most common problems of distraction and disorganization.

It's not a substitute for everything you've read in this book. But in a pinch, these specific solutions to the most common problems—provided by our coauthor, Coach Meg—will help.

1. ABSENTMINDEDNESS
("Where did I leave my keys?")

When we lose our *mindfulness,* meaning our complete presence and attention to the task at hand, it feels like we're losing our minds. But there's a difference between your mind and your mindfulness! When we're in the shower, we're thinking about a conflict with a work colleague. When we're driving a car, we're thinking about how our kids will be upset by being home late. When we're eating a delicious meal, we're worried about paying bills. When we set down our keys or park a car in a parking garage, we're already on to the next task and don't notice where we left them.

It wasn't always that way. We were good at being mindful when we were children. As adults, our minds fill up with stresses, strains and huge responsibilities as we grow more mature. There are many routes to recultivating that mindfulness we once had. Here's one path to follow that will also lead you back to your car keys!

- *Build awareness:* Start noticing when you are mindful or you've lost mindfulness—jot it down in a notebook or create a notes page on your cell phone.

- *Set a goal:* Think about what percent of the time that you are mindful and totally awake and present in the moment. Perhaps you're at a 5 out of 10. Where do you want to get to and by when?

- *Shift sets to gratitude in an instant:* The present is a gift. Appreciate it. Or stop and take a few deep breaths and focus on just your breathing to slow down your racing mind.

- *Practice becoming mindful:* Concentrate on one small task at a time for one week. For example, this week I will be mindful about where I put my cell phone. I will take a few seconds to consider where to store it in my handbag or pocket and notice where I'm placing it. Next week I'll work on my keys. The week after, my glasses. Then I'll fully enjoy my first cup of coffee or savor a small piece of dark chocolate in the late afternoon.

2. EASILY DISTRACTED
("Oh, that looks interesting...")

We've become a society of instant gratification. We can't wait. We want to know. We *need* to know. Right now! Sometimes this training is just what we need so that we are responsive to urgent and valid needs and requests. Unfortunately, most new information isn't urgent and perhaps not even important. However, we haven't trained our brains to handle the second step: a quick set-shift to ask, "Is this urgent?" If not, return quickly and effectively to your current focus. If so, do you need to rejigger your priorities or not?

To do this, you need to develop a two-step brain pattern.

Step 1: Evalute.

Step 2: Shift back to the present focus if the information isn't urgent, and jump off and focus on the new information if you deem it high priority.

A helpful way to do this is to rate the urgency and importance of each new message or input. Give it a rating of 1–10. Anything 7 and over demands immediate attention, so you'll need to set shift from what you're doing to this new information. A rating of 4–6 depends on the task at hand. If this new message or stimuli pops up at a time when you're not feeling swamped or busy with something else, attend to it. If not, respond later. Anything below 4, you can probably ignore for the time being. (Remember here that technology can help us in this process: most cell phones now ask us if we want to listen to a new call or read a new text now or later; with e-mail you have the option to click on the pop-up announcing a new message or not.)

3. OVERWHELMED BY CLUTTER
("Ugh, this place is a mess...")

It's amazing how clutter can impact our brains, making our minds feel as cluttered as our sock drawer. Don't you envy the people who can tolerate, even enjoy, chaotic messes and seem immune to this effect?

Decluttering your life requires a long-term plan—at least three months, possibly up to one year. Take the small, gradual measures needed to untangle the messes and restore order, both outside (your home/office) and inside (your brain). But fear not, follow these steps and the end of clutter will be in sight! (Along with the bottom of your desk that probably you haven't seen in months.)

Also, while it may seem that the onus is on you and you alone to clean up the mess you've made, this is where we can tap into your undiscovered forces.

- Getting a partner—your child, mate or friend—to help here can really help declutter. They offer a fresh perspective; they can see the sky from the messy weeds. Invite them to help, be open to their suggestions about what should go where and make it fun. Celebrate a few hours of decluttering by enjoying a great meal (your treat!) or going for a long walk to celebrate your relationship and productive time together. You could find a partner who also has decluttering challenges and alternate spaces.

- After a session with your partner, schedule solo decluttering periods of an hour, once a week at first and eventually once a month, to make sure you stay on top of this.

- Schedule ongoing decluttering time—fifteen minutes per week to keep decluttered areas under control.

- Be sure to focus on and appreciate how many areas have been brought to order along the way, not on how much is left to do.

4. CAN'T FOCUS ("Okay, come on, I've really got to pay attention to this now...")

Recall the last time you *were* able to focus on one thing 100 percent—a movie, a book, a sports event, a riveting or important presentation or a doctor's appointment. Think about that time. Is it just that you have forgotten how to focus and need some practice in reclaiming that knowledge? Or have you never had the experience of true focus? Either way, we can help.

If you can remember a time when you could focus, recall the conditions—where were you, what time of day or week and what were the conditions that enabled this focus? What are the conditions under which focus can flourish? No doubt the saliency or interest of what you were focusing on was a big part of it. You can't control that, but you *can* recreate the conditions surrounding it.

Think: perhaps it was in the morning when you were fresh. Maybe it was after a good night's sleep, a fun evening with friends, an intimate connection with your mate or a nutritious breakfast? Come up with the top three things that helped you focus on that day of "full focus," and start experimenting. Try to recreate or at least reimagine those conditions before your next important task or moment.

If you can't recall a time when you could focus—then you need to start now by becoming very deliberate and mindful. Say to yourself, "Now is the moment I want to focus and learn this skill, just as an infant brings his entire attention to getting off of all fours and up on two legs." Take some deep breaths and clear the noise in your mind. Then work on focusing on a task—a monthly report, a memo, preparations for an important meeting—for five minutes. Stop at five minutes (unless you lose focus earlier) and clap like we do for a child who takes his first steps (or at least give yourself a pat on the back). You made it to five minutes of full-throttle focus. Fabulous!

Now go for ten. Notice what's working and what isn't. Like walking or riding a bike, it's about practice and your willingness to fall down and get back up again.

5. CHRONIC LATENESS
("Uh, sorry...")

This may be a sign that your commitments are beyond your personal bandwidth. You may have simply exceeded what you can carry. If so, consider the following suggestions:

- *Trim your sails:* Write down a list of your commitments—daily, weekly, monthly. (Your spouse or a partner can help.) Determine if some of these can be jettisoned, delegated or trimmed down. In that way, try and reduce your list of regular commitments by at least 10 percent. Better to do fewer things well then many things poorly.

- *Get your fifteen-minute daily down time:* Lateness and forgetfulness may be a sign that you need some downtime to restore calm and balance and increase brain function. Harvard University mind and body expert Herb Benson recommends ten to fifteen minutes a day of a repetitive, mindful activity (deep breathing, meditation, yoga). You can do it in the morning to start the day on a calm footing or in late afternoon before the evening starts.

- *Adjust your emotional balance:* This may be a sign of too few positive emotions or too many negative emotions, both of which hurt brain function, particularly memory. Check your ratio of positive to negative emotions at www.positivityratio.com. The tipping point is 3:1 above which our brains function well and below which our brains don't function well at all.

6. CAN'T KEEP ALL THE BALLS IN THE AIR
("I thought I was doing it all...")

If by this, the concern is that you can't multitask, don't fret. Despite the conventional wisdom, research has shown that multitasking isn't very effective anyway.

Each task, brief or otherwise, is best done with your full attention, not a quarter, half or even three quarters of your attention. The new skill to learn is to bring your entire consciousness to each task, whether it's talking to your kids, answering an e-mail, even looking out the window to appreciate something pretty. Imagine it like turning your head and fixing the gaze of another and connecting fully, as we do when we are falling in love and want to send a sign of our feelings. You need to make a clear break, a mental transition from task to task and not let the previous task or future task infect the current one. When you bring your full presence to a task, time slows down and expands and much can be done even in moments.

This is about attending beautifully to each thing, not about getting lots of things done half-baked.

7. TREADING WATER
("I am just hanging on here...")

How do you avoid the sense that you can't get ahead because you're just trying to keep up with the constant wave of demands on your time? To stop treading water, and start moving forward, you need a sense of greater control. This will help give you confidence that the things you need to get done *will* get done as well as the peace of mind that comes with knowing you are moving in the right direction.

To regain these important qualities, I recommend you go on the Time-Zone Diet! Here's what I'm talking about:

- Schedule interruption-free zones, the most productive times each day. Start with fifteen minutes, then thirty minutes and build up to several hours a day over a few months.

- Schedule zones to deal with interruptions when you need a break from demanding projects—say twenty minutes per day—to check texts, calls, tweets, etc.

- Let people who text, tweet, call or e-mail know that you have scheduled zones for responding and you will not respond immediately.

- Practice not looking at texts, e-mails and such when you are in an interruption-free zone. You'll quickly enjoy the sense of control that comes from not responding in a knee-jerk fashion.

- Take a five-minute break every ninety minutes so that you are more attentive and feel in control of your calendar and your day.

8. STRESSED OUT
("This is all just too much...")

Granted, who isn't stressed out these days? But this is the kind of stress that is often caused by distraction and disorganization and is leading to an overall deterioration of physical and mental health.

"I can't find the time to get to the gym, I'm too crazed to take the time to prepare meals so I grab fast food, and I'm so frazzled that I can't sleep. I'm a stressed-out wreck!"

We hear this a lot, and while our clients don't say it in so many words, it's a cry not necessarily of depression but of disorganization and distraction. When you get to this point, your check-engine light is flashing brightly. Time to stop. Time to hit the reset button. How can you shift from being a wreck to being someone people look up to as you appear to effortlessly glide through your day—fit, healthy, well-nourished and well-rested?

- First, take a few moments a few times a week to review what you've got going, the good things in your life that you should be grateful about. This may sound trite, but it will help you shift to a more positive footing.

- Second, redefine who you want to be—moving from stressed-out wreck to what? Calm and confident? Relaxed and at peace? Who's your role model? How would you describe that person?

- Turn priorities upside down and take care of your health first, which will give you more energy, balance and calm to get done what needs to get done.

- Find one health behavior to get under control—just one. Maybe it's exercise and perhaps exercising somewhere more convenient than the gym will take less time. So get out and walk or buy a workout video you can do at home. You'll be doing something good for your health *and* your stress levels!

9. EFFICIENT, BUT COULD BE MORE SO ("I could do so much more...")

You say you've got the big things under control? You seem to have things prioritized and running smoothly in your life—but you still feel you could be more efficient?

First, congratulate yourself on the fact that you don't feel like you're coming apart at the seams, as so many people do these days. Now, let's look more closely at what you're trying to do: what does *efficient* mean? The dictionary describes it as performing or functioning in the best possible manner with the least waste of time and effort. So where are you losing time in your day? There's probably not much "fat" left to trim, but perhaps you can find some small things. Let's look at some possible ways to reduce time/effort waste:

- Start the day by building a to-do list: create your daily list the evening before so that you start the day with the list of key things that must get done.

- Put out work clothes, workout clothes, snack bag, handbag (with keys and glasses); recharge phone and backup computer the evening before.

- Go through your calendar or to-do list, and take a hard look at all the things you're doing (journaling can help here as well). Chances are you'll find a number of things you're doing that could be trimmed from your day.

- Take short breaks (every ninety minutes) to recharge rather than processing tasks inefficiently because you've got minor burnout.

- Schedule periods for creative and demanding projects and intersperse with quick, handle-turning projects to give your brain a rest. Don't waste your peak energy slots on the little things.

10. ON THE BRINK OF CHAOS
("#@?!!*...")

A lot of people today have this nagging sense that they are teetering on the edge and that all it will take is one more assignment or one more new technology they have to master to send them plummeting into an abyss of chaos.

If you feel that way, remember this: while it may seem that you're about to go over the edge, you haven't—at least not yet. The thing to deal with is your "nagging sense." You are choosing to have negative thoughts that nag you. Perhaps you could jump out of the nagging-thought pattern in your brain and find another thought pattern; one like "I'm doing a good job; I'm keeping my cool and balance in face of the risk of going over the edge, like a sure-footed mountain goat perched on a cliff edge."

So shift your mental picture from one of you trembling on the edge of the precipice to you confidently walking along the edge, filled with the promise of discovery. Yes, that's you, on your way to great new things. Enjoy the view!

NOTES

1. Dr. David Lewis. "Psychologically Toxic Office Space." The Esselte Corporation, October 18, 2005. http://corporate.esselte.com/enUS/PressReleases/Psychologically_Toxic_Office_Space.html

2. Katherine Vessenes. "Advisers Can Work Less, Be Healthy and Make More Money." Financial Planning Association, 2008. http://www.vestmentadvisors.com/services/research/press-report.pdf

3. American Psychological Association. "Stress a Major Health Problem in The U.S., Warns APA." October 24, 2007. http://www.apa.org/news/press/releases/2007/10/stress.aspx

4. Mike Maseda. "Healthy, Stress-Free Workplace Benefits Employees, Bottom Line." *Houston Business Journal,* September 5, 2004. http://www.bizjournals.com/houston/stories/2004/09/06/focus5.html

5. Bill Breen. "Desire: Connecting with what Customers Want." *Fast Company*, January 31, 2003. http://www.fastcompany.com/magazine/67/desire.html?page=0%2C0

6. Society for Human Resource Management. "Employees Admit Self-Imposed Pressure to Work Long Hours," May 12, 2009. http://www.shrm.org/about/pressroom/PressReleases/Pages/SHRMPollEmployeesAdmitSelf-ImposedPressuretoWorkLongHours.aspx

7. David L. Strayer, Frank A. Drews, Dennis J. Crouch and William A. Johnston. "Why Do Cell Phone Conversations Interfere With Driving?" University of Utah, Department of Psychology, June 29, 2006. http://www.psych.utah.edu/AppliedCognitionLab/CogTechChapter.pdf

8. Marcel Adam Just, Timothy A. Keller and Jacquelyn Cynkar. "A Decrease in Brain Activation Associated with Driving when Listening to Someone Speak." Center for Cognitive Brain Imaging, Department of Psychology, Carnegie Mellon University, December 28, 2007. http://www.ccbi.cmu.edu/reprints/Just_Brain-Research-2008_driving-listening_reprint.pdf

9. National Highway Traffic Safety Administration. "Distracted Driving 2009." U.S. Department of Transportation, September 2010. http://www.distraction.gov/research/PDF-Files/Distracted-Driving-2009.pdf

10. S.P. McEvoy, M.R. Stevenson, A.T. McCartt, M. Woodward, C. Haworth, P. Palamara and R. Cercarelli. "Role of Mobile Phones in Motor Vehicle Crashes Resulting in Hospital Attendance: a case-crossover study." *British Medical Journal,* 2005. http://www.ncbi.nlm.nih.gov/pmc/articles/PMC1188107/?tool=pubmed

REFERENCES

INTRODUCTION

Page viii

Wager TD, et al. "Prefrontal-Subcortical Pathways Mediating Successful Emotion Regulation." *Neuron* 59 no.66 (Sept 2008): 1037-50.

Page xiii

Institute for the Future. "Sensory Transformation: New Tools & Practices for Overcoming Cognitive Overload." May 2007. www.iftf.org.

Page xvi–xvii

U.S. Department of Transportation. Official Government Website for Distracted Driving. www.distraction.gov.

Page xvii

Pew Internet & American Life Project, Pew Research Center. "Teens and Mobile Phones." April 2010. www.pewinternet.org.

CHAPTER 1

Pages 4–5

Kessler RC, et al. "The World Health Organization Adult ADHD Self-Report Scale (ASRS): A Short Screening Scale for use in the General Population." *Psychological Medicine* 35, no.2 (Feb 2005): 245-56.

Page 11

The National Institute of Mental Health Strategic Plan, 2008.

Page 11

Mitchell JP, et al. "Media Prefrontal Cortex Predicts Intertemporal Choice." *Journal of Cognitive Neuroscience* 23, no.4 (Apr 2011): 857-66.

CHAPTER 2

Page 33

Cohn MA, et al. "Happiness Unpacked: Positive Emotions Increase Life Satisfaction by Building Resilience." *Emotion* 9, no.3 (Jun 2009): 361-8.

CHAPTER 3

Page 50

Whalen PJ and EA Phelps, eds., *The Human Amygdala.* New York: Guilford, 2009.

Page 51

Blair KS, et al. "Modulation of Emotion by Cognition and Cognition by Emotion." *NeuroImage* 35, no.1 (Mar 2007): 430-40.

Page 53

Wager TD, et al. "Prefrontal-Subcortical Pathways Mediating Successful Emotion Regulation." *Neuron* 59, no.66 (Sept 2008): 1037-50.

Page 56

Johnstone T, et al. "Failure to Regulate: Counterproductive Recruitment of Top-down Prefrontal-subcortical Circuitry in Major Depression." *Journal of Neuroscience* 27, no.33 (Aug 2007): 8877-84.

Page 57

Van Dillen L, et al. "Tuning Down the Emotional Brain: An fMRI Study of the Effects of Cognitive Load on the Processing of Affective Images." *NeuroImage* 45, no.4 (May 2009): 1212-19.

Pages 60–61

Denson, et al. "The Angry Brain: Neural Correlates of Anger, Angry Rumination and Aggressive Personality J. Cog." *Neuroscience* 21, no.4 (Apr 2009): 734-44.

Page 61

Hooker CI, et al. "Neural Activity to a Partner's Facial Expression Predicts Self-regulation After Conflict." *Biological Psychiatry* 67, no.5 (Mar 2010): 406-13.

CHAPTER 4

Pages 87–88
Csikszentmihalyi M. *Flow: The Psychology of Optimal Experience.*
New York: Harper Perennial, 1990.

Page 90
Linley A, et al. *The Strengths Book: Be Confident, Be Successful, and Enjoy Better Relationships by Realising the Best of You.* Coventry: CAPP Press, 2010.

CHAPTER 5

Page 99
Barkley RA "Attention Deficit Hyperactivity Disorder." In Mash EJ & Barkley RA, eds., *Child Psychopathology.* 2nd ed. New York: Guilford, 2003.

Page 102
Boonstra AM, et al. "To Act or Not to Act, That's the Problem: Primarily Inhibition Difficulties in Adult ADHD." *Neuropsychology* 24, no.2 (Mar 2010): 209–21.

Page 102
Goos LM, et al. "Validation and Extension of the Endophenotype Model in ADHD Patterns of Inheritance in a Family Study of Inhibitory Control." *American Journal of Psychiatry* 166, no.6 (2009): 711–17.

Pages 106–107
Gray R. "A Model of Motor Inhibition for a Complex Skill: Baseball Batting." *Journal of Experimental Psychology* 15, no.2 (Jun 2009): 91–105.

CHAPTER 6

Page 128
Bays PM & Husain M. "Dynamic Shifts of Limited Working Memory Resources in Human Vision." *Science* 321, no.5890 (Aug 2008): 851–4.

Page 129
McVay JC & Kane MJ. "Conducting the Train of Thought: Working Memory Capacity, Goal Neglect, and Mind Wandering in an Executive-control Task." *Journal of Experimental Psychology* 35, no.1 (Jan 2009): 196–204.

Page 130

Charlton RA, et al. "White Matter Pathways Associated with Working Memory in Normal Aging." *Cortex* 46, no.4 (Apr 2010): 474–89.

Page 130

Takeuchi H, et al. "Training of Working Memory Impacts Structural Connectivity." *The Journal of Neuroscience* 30, no.9 (Mar 2010): 3297–303.

Page 138

Erickson KI, et al. "Aerobic Fitness is Associated with Hippocampal Volume in Elderly Humans." *Hippocampus* 19, no.10 (Oct 2009): 1030–9.

CHAPTER 7

Pages 147–148

Eling P, et al. "On the Historical and Conceptual Background of the Wisconsin Card Sorting Test." *Brain and Cognition* 67, no.3 (Aug 2008): 247–253.

Page 148

Hedden T & Gabrieli J. "Shared and Selective Neural Correlates of Inhibition, Facilitation, and Shifting Processes during Executive Control." *NeuroImage* 51 (May 2010): 421–31.

Page 149

Perry ME, et al. "White Matter Tracts Associated with Set-shifting in Healthy Aging." *Neuropsychologia* 47, no.13 (Nov 2009): 2835–42.

Page 152

Burgess PW, et al. "The Cognitive and Neuroanatomical Correlates of Multitasking." *Neuropsychologia* 38, no.6 (Jun 2000): 848–63.

CHAPTER 8

Page 168

National Institute of Mental Health press release. "$40 Million Awarded to Trace Human Brain's Connections." September 15, 2010. www.nih.gov.

INDEX

ACKNOWLEDGMENTS

We wish to thank the following people:

Julie K. Silver, MD, Chief Editor of Books at Harvard Health Publishing, the organizing force behind *Organize Your Mind, Organize Your Life,* for bringing this team together, and providing the inspiration and oversight needed to get us to the finish line.

Linda Konner, our wise literary agent, for her insight and unique prowess in navigating books from concept to completion.

Deb Brody, our editor, for her foresight and enthusiasm for this project, and for making the editing process, often a painful one, a joy.

The brilliant thinkers, teachers, researchers, coaches—some of them cited in the following pages and in the Appendix—whose work provided the scientific bedrock of this book. For their special insights on working memory and how to improve it, we thank Marie Pasinski, MD, a neurologist at Harvard Medical School and author of *Beautiful Brain, Beautiful You,* and Martha Wolf, director of the Alzheimer Center at Parker Jewish Institute for Health Care and Rehabilitation in New Hyde Park, New York.

Robert Shmerling, MD and Catherine Smith for allowing us to peek into their organized lives.

Above all, for the innumerable ways in which they enrich our lives, our colleagues, mentors, patients, clients and students at the various organizations we are proud to represent: Harvard Medical School,

Massachusetts General Hospital and McLean Hospital Department of Psychiatry, Institute of Coaching at McLean Hospital/Harvard Medical School, New York Institute of Technology, and Wellcoaches.

And finally, our families and friends for their love and support.

PAUL HAMMERNESS, MD is an assistant professor of psychiatry, Harvard Medical School; assistant psychiatrist, Department of Psychiatry, Massachusetts General Hospital; and child and adolescent psychiatrist, Newton Wellesley Hospital. Dr. Hammerness has been involved in research on the brain and behavior for the past 10 years, with a focus on Attention Deficit Hyperactivity Disorder (ADHD). He has lectured on the topic locally, nationally and internationally to other physicians, mental health professionals, educators and families. Dr. Hammerness is an active clinician as well, treating children, adolescents and adults in his clinical practice.

MARGARET MOORE, aka Coach Meg, is the founder and CEO of Wellcoaches Corporation (www.wellcoaches.com), a leader in building international standards for professional coaches in health and wellness. Margaret is codirector of the Institute of Coaching at McLean Hospital, an affiliate of Harvard Medical School; a founding advisor of the Institute of Lifestyle Medicine at Harvard Medical School and lead author of the first coaching textbook in health care (*Coaching Psychology Manual,* published by Lippincott).

In addition to coaching coaches, Coach Meg has guided thousands of coaches and hundreds of clients through the change and growth process that we describe in this book. Her strength is the translation of evidence-based theories and concepts into simple and practical approaches that catalyze lasting change and transformation, changing the world one person at a time.

She blogs on coaching and change for *The Huffington Post* and *Psychology Today.*

JOHN HANC teaches writing and journalism at the New York Institute of Technology. A longtime contributor to *Newsday* and a contributing editor to *Runner's World* magazine, his work also appears in *The New York Times, AARP Bulletin, Family Circle, Smithsonian* and *Yoga Journal.* In addition, Hanc is the author of eight books. His most recent, *The Coolest Race on Earth* (Chicago Review Press, January 2009), a memoir of his experiences running the 2005 Antarctica Marathon, won an award at the 2010 American Society of Journalists and Authors' national writing competition.